Collins

AQA GCSE 9-1 Revision

Geography

Geography

AQA
GCSE 9-1

Revision
Guide

Paul Berry, Brendan Conway, Janet Hutson, Dan Major and Robert Morris

Contents

Contents

Geographical Skills

You must be able to:

- Demonstrate a range of geographical skills including cartographic, graphical, numerical and statistical.

Cartographic Skills

Latitude and Longitude

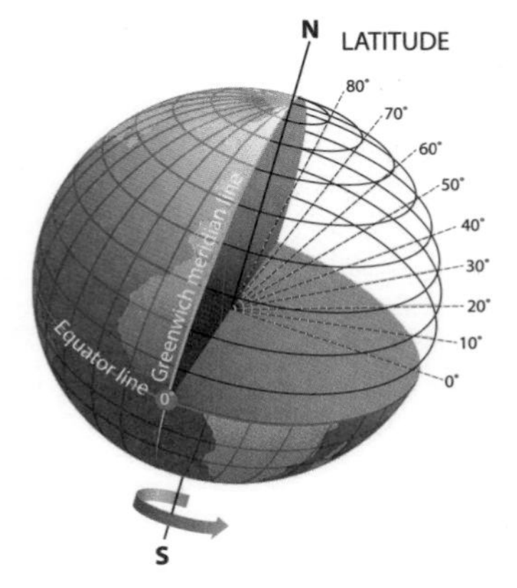

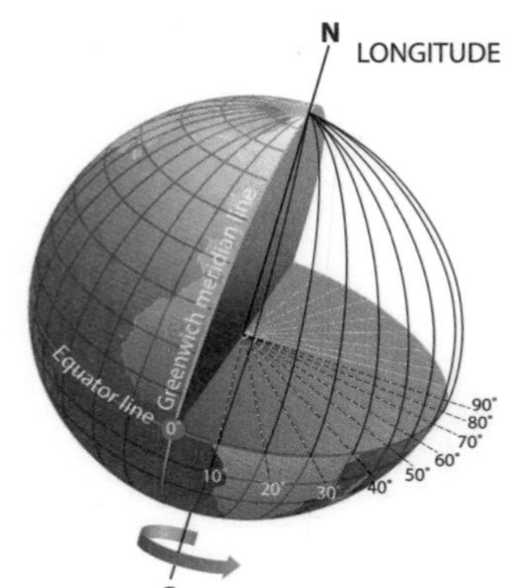

- All points on Earth can be located precisely with coordinates using **latitude** and **longitude**.
- Latitude lines provide a measure of how far places are north or south of the Equator. Latitude lines are all parallel to the Equator.
- Longitude lines provide a measure of how far places are east or west from the Prime Meridian (sometimes called the Greenwich Meridian). Longitude lines run between the North Pole and South Pole (meeting at the Poles) at right angles to the latitude lines.

Atlas Maps

- Categories of map include:
  - General: physical (showing **relief**, drainage), human (showing population distribution, population movement, transport networks, settlement layout), and **political** (showing borders).
  - Thematic: weather; climate; ecosystems; demographic (population); development; economic (industry).
  - Environmental issues: climate change; pollution; desertification; deforestation.

Patterns and Distributions

- You will need to be able to recognise:
  - Patterns: where are things concentrated in **dense** clusters or more **sparse**?
  - Patterns found in: land use; populations; settlements; communication networks (transport and telecommunications); earthquakes; tropical storms.
  - Patterns that have recognisable shapes, e.g. **nucleated**, **linear**, **dispersed** or **radial**.
  - Distributions: describe where things are in a broader sense, using compass points and terms such as central, coastal and **peripheral**.

Ordnance Survey (OS) Maps

- You should be able to locate places using grid lines: give the number of the horizontal **easting line** first, then the vertical **northing line**. A rule may be useful, such as 'along the corridor and up the stairs'.

- Use four-figure grid references to locate whole grid squares – this is best for larger areas.
- Use six-figure grid references to locate specific points.

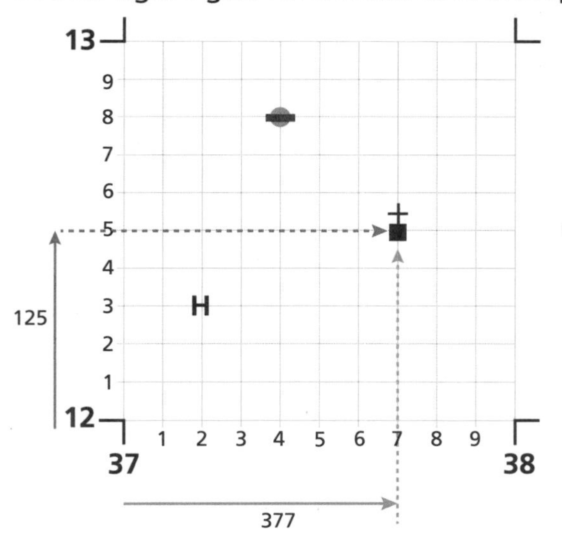

The six-figure grid reference for the church is:

377125

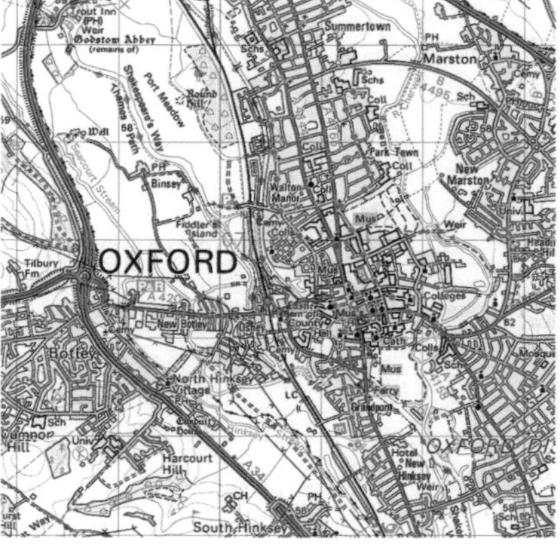

Ordnance Survey mapping

Scale

- All maps use scale to provide a standardised way of representing the real world on a page or device screen. Map scales often use a : (colon) format, which means 'represents' or 'equal to in reality'. Common OS map scales are 1:25 000 and 1:50 000.
- Use the scale to check your technique for measuring straight- and curved-line distances on maps.

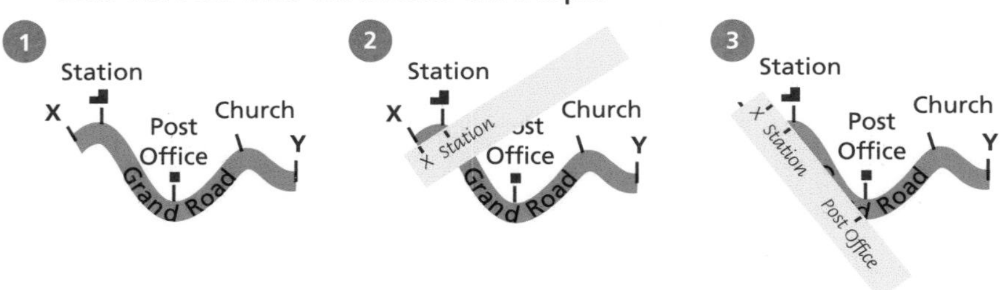

Compass Directions

- Learn the points of the eight-point compass and practise using them to describe locations, e.g. 'the camera is pointing south-west'; 'the river flows in an easterly direction'.

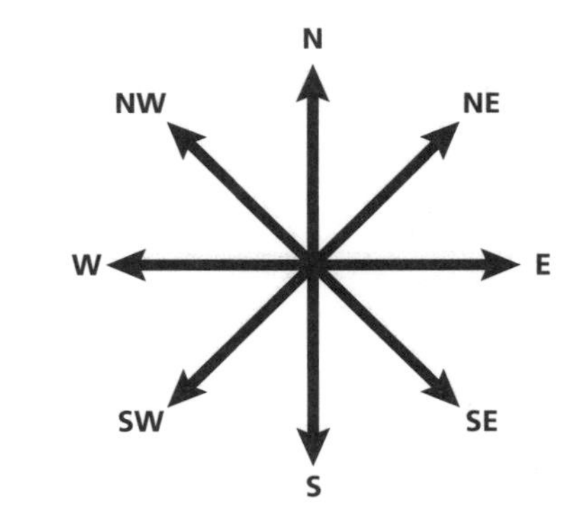

Isolines as Used on Relief and Meteorological Maps

- Relief is the shape or topography of the land, shown by isolines called contours or by layer colouring.
- Height is usually measured in metres above sea level.
- Points of exact height are shown using spot heights or triangulation pillars.
- Contours are lines along which the height is the same.
- Steep slopes are shown by closely spaced contours; gentle slopes are shown by contours with larger spaces between them.
- Landscape shapes include valleys, ridges, spurs and plateaus.
- Descriptions can also refer to their shape, size, elevation and compass direction (orientation).
- Cross-sections are like slices through a landscape, drawn using contours. These can be drawn by hand or with GIS **(geographical information system)** elevation profiles.

Contour Patterns and Corresponding Relief Diagrams

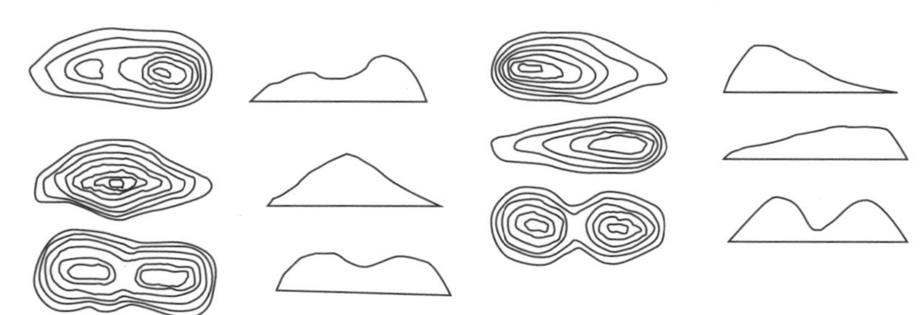

- Helpful terms to describe hills or valleys include **symmetrical** (similar gradient on both sides) or **asymmetrical** (steeper on one side than the other).
- Relief has a significant impact on human activity, e.g. land use and communications.
- **Isolines** are often used to show weather data such as pressure (**isobars**) and rainfall (**isohyets**).
- They can also be an effective way to map data such as pedestrian counts or pollutants.

Map Symbols

- Always refer to the symbology of maps, using the key provided.
- Practise interpreting by making links between symbols or their absence. For example, on an OS map, a lack of human activity on land beside a river (where there are large gaps between contours) could be due to a desire to avoid flooding.

Drainage: Water on the Land

- River flows or an absence of rivers tell us about the drainage of water from a landscape.
- Drainage density is measured by the total length of river channels in an area, e.g. km per km^2, and is usually high in areas with **impermeable** rocks and low in areas with **permeable** rocks.
- On some rivers channels, **hard** or **soft river engineering** is used as a form of **flood management**.
- Drainage includes lakes, waterfalls, underground rivers (e.g. in limestone areas) and artificial water features such as reservoirs.
- Terms used to describe drainage patterns include dendritic, rectangular, parallel, trellised, deranged and radial:

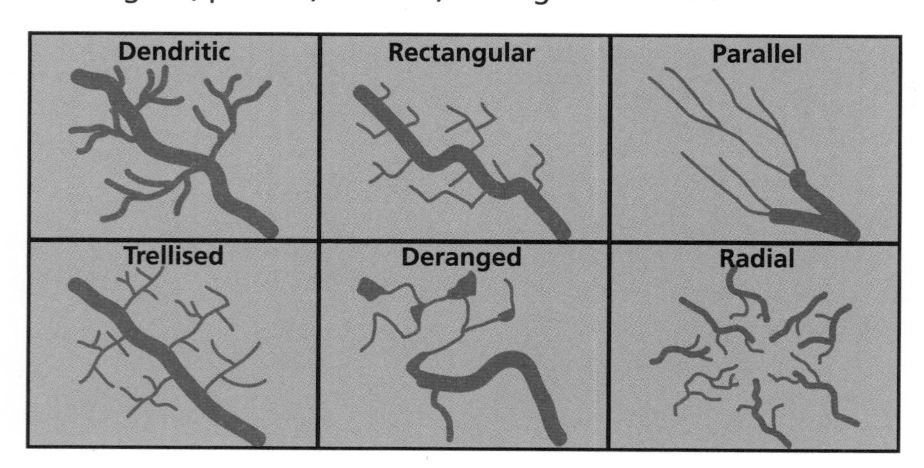

Building a Cross-section

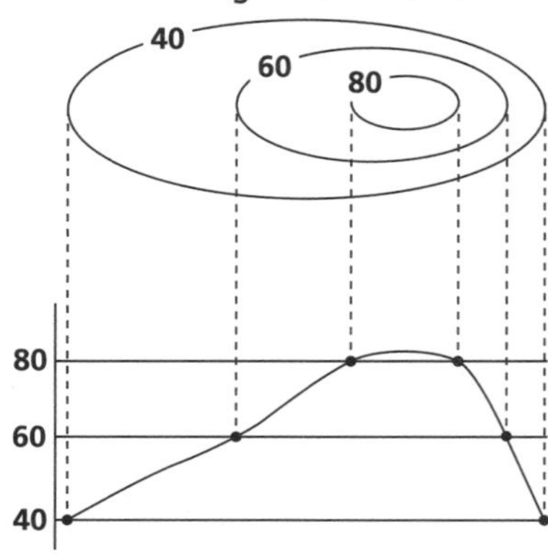

Imagine walking along a straight line from the west side of the hill to the east. You can draw a transect line to show your route. On the transect, put a small dot each time a contour is crossed and label it with its height. Use the line to plot the heights on graph paper or a chart. Then 'join the dots' to show the relief of the hill.

Key Point

You will use your geographical skills when answering questions about all the topics you have studied, and there are lots of opportunities to practise these skills throughout this book.

Key Point

Fundamental geographical skills include the ability to use and understand maps, graphs and charts, and being able to employ numerical data and statistical techniques effectively.

Graphical Skills

- You should be able to suggest appropriate forms of graphical representation. You should also be able to interpret and extract information from, and add to, the following:
  - scatter graphs; line charts; pie charts; bar charts; histograms with equal class intervals; particular types of charts such as population pyramids, divided and cumulative bar and line charts

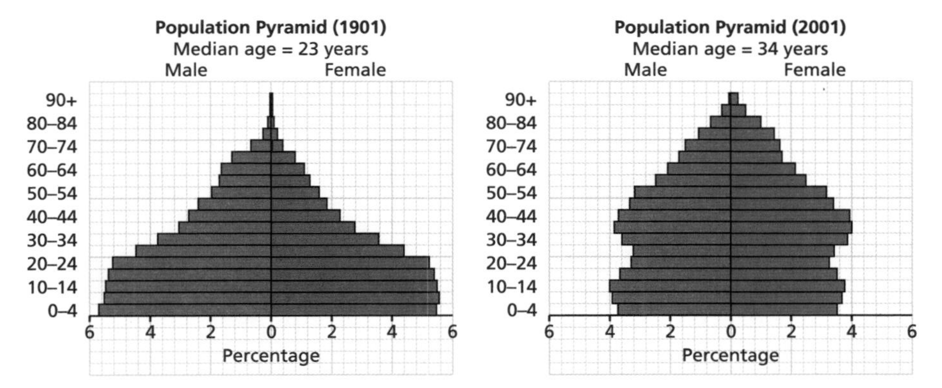

  - pictograms, proportional symbols, choropleth maps, isolines, dot maps, desire lines and flow-lines
  - aerial imagery; satellite imagery; false colour images.
- Fieldwork can often be enhanced by photographs and field sketches, which are really 'diagrams' of a place. Good annotations can add value to graphics.
- Sketch maps are roughly drawn maps that show basic details of an area, and can be labelled or annotated to identify key features.

Numerical Skills

- Number, area and scales; quantitative relationships between units.
- Data collection: accuracy; sample size and procedures; control groups; reliability.
- Proportion and ratio; magnitude; frequency.

Statistical Skills

- Measures of central tendency: median; mean; mode and modal class.
- Measures of spread and cumulative frequency: range; quartiles and interquartile range; dispersion graphs.
- Percentages: percentage increase or decrease; percentiles.
- Relationships in bivariate data: trend lines through scatter plots; lines of best fit; positive and negative correlation; strength of correlation.
- Predictions and trends: interpolation; extrapolation.
- Limitations and weaknesses in selective statistical presentation of data.
- Geospatial (geolocated or georeferenced) data presented in a GIS framework. GIS can also be used to analyse spatial data.

Key Words

latitude
longitude
relief map
political map
dense
sparse
nucleated
linear
dispersed
radial
peripheral
easting line
northing line
scale
contours
spot heights
triangulation pillars
valley
ridge
spur
plateau
orientation
GIS
elevation profiles
symmetrical
asymmetrical
isolines
isobars
isohyets
impermeable
permeable
hard river engineering
soft river engineering
flood management
histogram
choropleth map
quantitative
frequency
median
mean
mode
modal class
quartiles
percentiles
interpolation
extrapolation

Tectonic Hazards 1

You must be able to:

- Understand that natural hazards are the result of physical processes
- Understand how tectonic hazards pose a risk to people and property
- Identify ways that effects from tectonic hazards can be reduced.

Natural Hazards and Their Impacts

- A natural hazard is an extreme event that causes loss of life, severe damage to property or severe disruption to human activities.
- Natural hazards can have a social, economic and environmental impact on an area, and they are more severe in lower income countries (LICs) because there is:
  - more poor quality housing and poor healthcare
  - poor infrastructure so it is harder to reach affected people
  - less money to protect people (e.g. with earthquake-proof buildings) and less money for responses (e.g. providing food).

Earthquakes and Volcanoes

- The Earth's crust is divided into seven large plates and 12 smaller plates that sit on the mantle below. They move very slowly in different directions due to heat convection currents in the mantle:
  - Magma below the surface is heated by the core and rises to the crust, where it cools and sinks, to be reheated once more.
  - Friction caused by the rising currents means that the plates are pushed forward or dragged back (plate tectonics).
  - There are two types of crustal plate: oceanic plates (tend to be denser but thin) and continental plates (less dense and thicker).
  - Two (or more) plates meet at a plate boundary or **plate margin**.
- There are four main types of plate boundary:
  - **Destructive margins** (or convergent boundaries) usually involve an oceanic plate and a continental plate moving towards each other and colliding. The denser oceanic plate is forced beneath the less dense continental plate (a process known as **subduction**) and land is destroyed. The melting plate causes volcanic activity and the stress created by friction between the colliding plates can cause earthquakes.
  - **Constructive margins** (or divergent boundaries) are where plates pull apart, allowing magma to reach the surface and erupt through fissures and faults, creating volcanoes. Earthquake activity also occurs at constructive boundaries.
  - **Conservative margins** are where plates may move in opposite directions to each other, or in the same direction, but at different speeds. As the plates move, their edges stick together, causing friction and stress to build up until one plate jolts forward. This sudden release of energy can cause earthquakes and rift valleys. There is no volcanic activity at conservative margins.

Key Point

The Earth is made up of:

- inner **core**: solid iron and nickel, temperature around 5500°C
- outer core: liquid iron and nickel
- **mantle**: semi-molten rock (magma)
- **crust**: thin layer of solid rock.

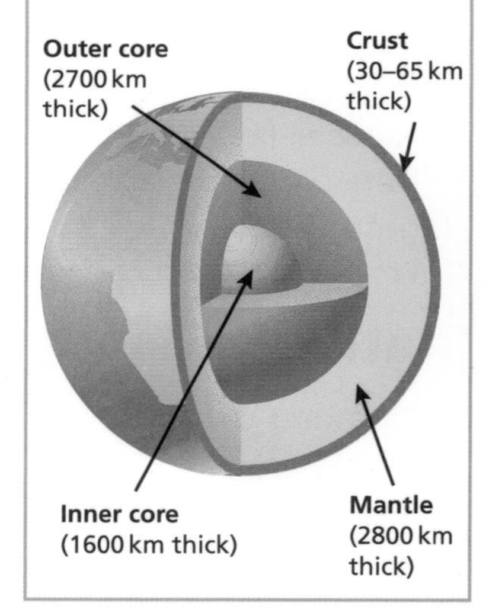

Key Point

Different physical processes occur at different types of plate boundary.

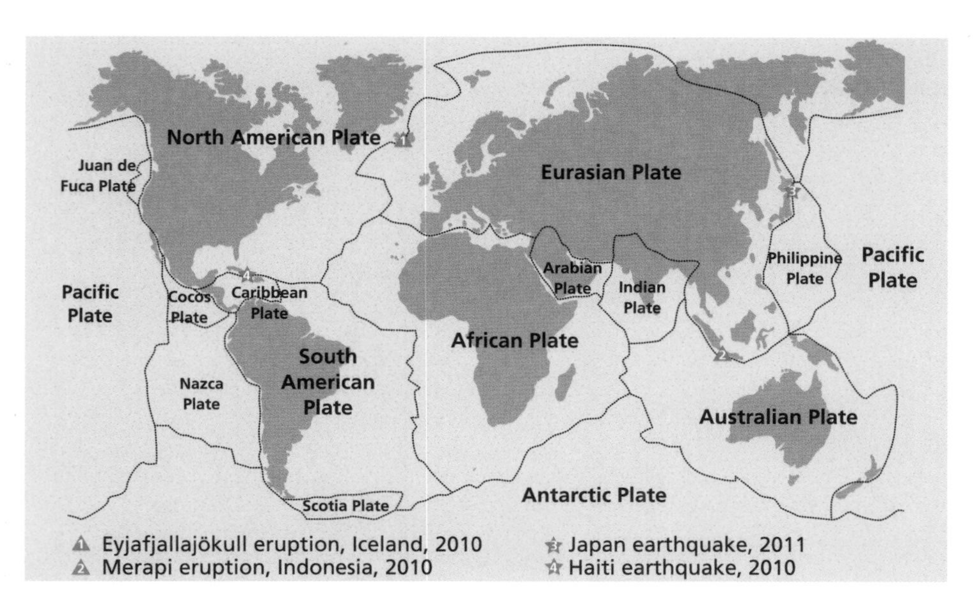

- Collision zones: if continental plates of the same density collide, they buckle up rather than subduct. This creates fold mountains. There is no volcanic activity at collision zones but earthquakes can occur, e.g. the 2015 Nepal earthquake.

- Most of the world's earthquake and volcano activity is found at the plate boundaries. 'Hot spots' are exceptions and can occur at any location where the mantle is very close to the surface.
- Volcanic activity is common where the plates meet around the rim of the Pacific Ocean; the area is known as the 'Pacific Ring of Fire'.

Why People Still Live in High-Risk Areas

- They are near friends and family and have a job there.
- Confidence that their government has the ability to 'fix' things.
- Attitude of 'it won't happen to me – it's safe'.
- Volcanic areas have fertile soils, valuable minerals, geothermal energy and tourism opportunities.

Reducing the Risk and Effects

- The 3 Ps for locations close to earthquake and volcano zones are:
  - **Plan/predict**: It is possible to predict where earthquakes will occur, but not when. It is possible to predict volcanic eruptions; tell-tale signs include mini-earthquakes, escaping gas and a change in the volcano's shape. Governments can plan evacuations when volcanoes show signs of eruption.
  - **Protect**: Buildings and bridges can be designed to withstand earthquakes and strengthened to withstand the weight of volcanic ash. Firebreaks can reduce the spread of fire.
  - **Prepare**: Emergency services can be trained and prepared. People can be taught how to react in an earthquake or an evacuation. Countries can receive emergency aid or longer-term aid to help rebuild infrastructure and buildings.

Quick Test

1. Describe destructive and constructive plate boundaries.
2. Give two reasons why people still live near active volcanoes.

Destructive, e.g. the Nazca Plate subducting beneath the South American Plate

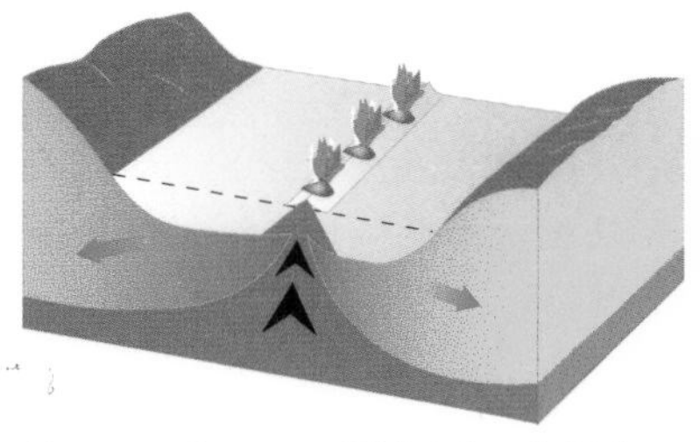
Constructive, e.g. Mid-Atlantic Ridge

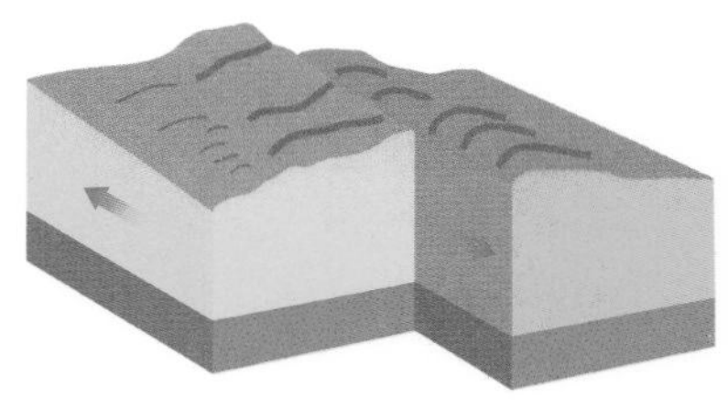
Conservative, e.g. San Andreas fault

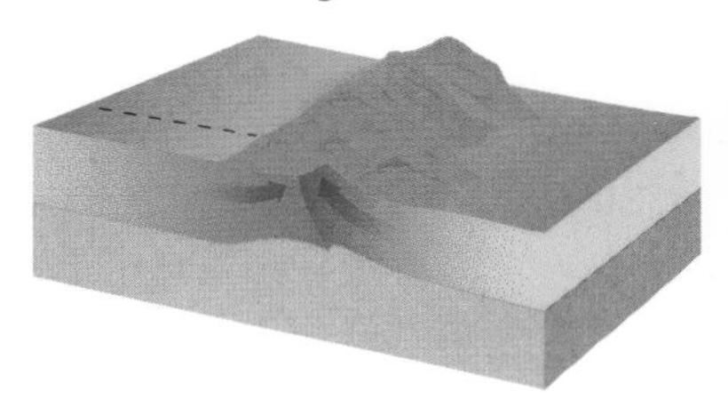
Collision zone, e.g. Himalayas and Alps

Key Point

Monitoring, prediction and planning can reduce the risk from tectonic hazards.

Key Words

core
mantle
crust
plate margin
destructive margin
constructive margin
conservative margin
collision zone
hot spot

Tectonic Hazards 2

You must be able to:

- Understand how the effects of and responses to earthquakes vary between areas of contrasting levels of wealth.

Anatomy of an Earthquake

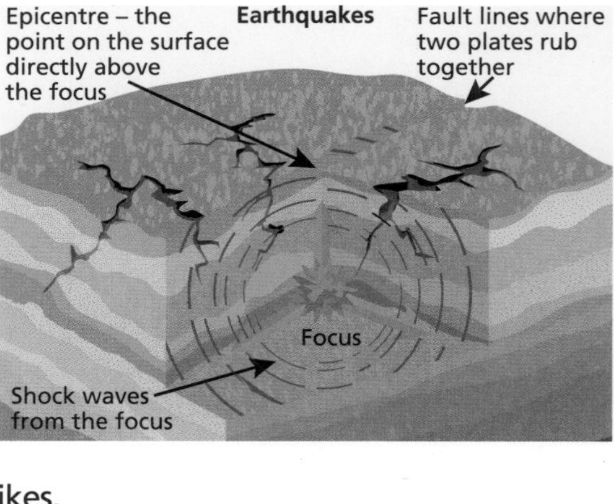

- Focus – point below surface where an earthquake occurs.
- **Epicentre** – the location on the surface directly above the focus.
- Seismic waves – shock waves of energy that travel through rock to the surface.
- Seismogram – measures shaking by an earthquake.
- Aftershocks – smaller tremors in the days after an earthquake.
- **Richter scale** – measures energy released in an earthquake.
- **Primary effects** – occur straight away when a tectonic hazard strikes.
- **Secondary effects** – occur later on, bringing more problems to those affected.

Earthquake Case Study – Tohoku, Japan, Higher Income Country (HIC)

Background

- 11 March, 2011 – 2:46 pm.
- Epicentre in the ocean 62 miles (100 km) north-east of Honshu island.
- A complex destructive plate margin involving the Pacific, Okhotsk, Eurasian and Philippine plates.
- 9.0 on the Richter scale.
- The focus was 18 miles (29 km) below the surface.
- Fourth largest earthquake in the world.

Primary Effects

- Buildings destroyed and whole towns demolished.
- Massive damage in Sendai City (1 million people).
- Homes destroyed.
- Loss of power supplies.
- Roads blocked.
- Water shortages.
- Effects felt in Tokyo 185 miles (298 km) away.
- Japan's main island (Honshu) moved 7.8 feet (2.4 metres).

Secondary Effects

- Resulting **tsunami** (with waves up to 40 m high) killed 15 000 people.
- 2 million people left homeless.
- Layers of mud and debris left on land.
- Buildings, vehicles and bodies washed out to sea.
- Workers unable to shut down the Fukushima nuclear reactor, allowing radiation to leak.
- Japan's stock market collapsed.

Aftermath of the 2011 Japan earthquake

Immediate Responses

- Aid came from many other nations.
- A nuclear exclusion zone was established.
- Tsunami warning issued 3 minutes after the earthquake.
- Japanese Red Cross mobilised emergency teams.
- Warning given to other locations, such as Hawaii.

Long-Term Responses

- New, efficient tsunami warning system.
- Modifications to tsunami walls and floodgates that had not been effective.

Key Point

Countries of contrasting levels of wealth show different effects and responses to tectonic hazards.

Earthquake Case Study – Haiti, Lower Income Country (LIC)

Background

- 12 January, 2010 – 4:53 pm.
- Epicentre 16 miles (25 km) south-west of Port-au-Prince city.
- 7.0 on the Richter scale.
- Focus only 10–15 km below surface.
- Haiti is the poorest country in the Western Hemisphere.
- North American Plate slid past the Caribbean Plate.

Primary Effects

- 316 000 killed.
- 1 million homeless.
- Difficult to get aid into the country.
- Homelessness, loss of power and roads blocked.
- Airport, port and hospitals closed.

Secondary Effects

- 1.6 million people in refugee camps.
- Water and food shortages.
- Increase in crime, particularly looting.
- Outbreaks of cholera.

One of the tent cities in Port-au-Prince following the 2010 earthquake

Immediate Responses

- Aid slow to arrive due to the damaged port and airport.
- The USA sent troops to support an aid programme.
- Many left Port-au-Prince city as their homes had been destroyed.

Long-Term Responses

- Dependence on overseas aid.
- $100 million aid from the USA and $330 million from Europe.
- New homes were built to a higher standard.
- Rebuilding of the port.

Key Words

epicentre
Richter scale
primary effects
secondary effects
HIC
tsunami
LIC

Quick Test

1. What is the difference between the primary and secondary effects of a tectonic hazard?
2. What is the difference between the focus and the epicentre of an earthquake?

Tectonic Hazards 3

You must be able to:

- Understand how the effects of and responses to volcanoes vary between areas of contrasting levels of wealth.

Features of a Volcano

- **Composite volcanoes** are made of alternating layers of lava and ash and are the result of multiple eruptions over hundreds of years. They give explosive eruptions of lava and ash. They are more likely to be found along destructive plate boundaries.
- **Shield volcanoes** form from runny magma (does not trap gases):
  - There is no build-up of pressure and eruptions are not explosive.
  - The runny lava flows a long way from the eruption, creating wide volcanoes.
  - They are more likely to be found on constructive plate boundaries, and at hot spots.
- **Pyroclastic flow** – torrent of hot ash, rocks, gases and steam, moving at up to 450 mph (700 km/h).
- Ash cloud – blocks out the Sun.
- **Lahar** – 60 mph mudslide of melted snow and volcanic ash.
- **Active volcano** – erupted recently.
- **Dormant volcano** – currently inactive but might erupt in future.
- **Extinct volcano** – not erupted for many thousands of years.

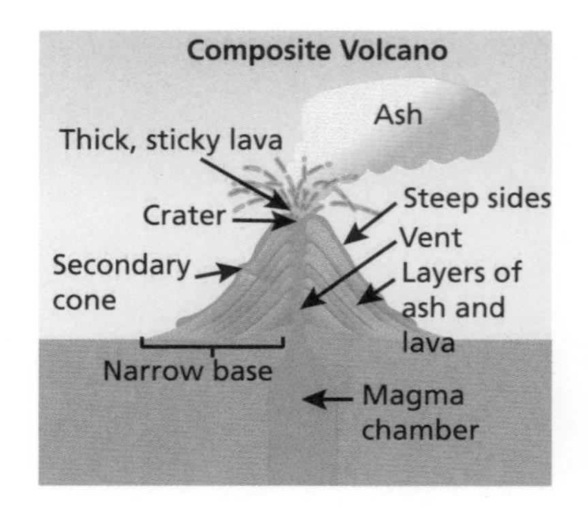

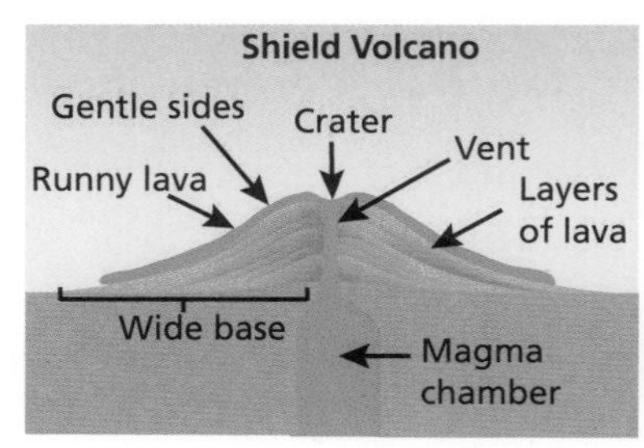

Volcano Case Study – Eyjafjallajokull, Iceland, Higher Income Country (HIC)

Background

- Erupted in April, 2010.
- Occurred at the spreading Mid-Atlantic Ridge of the constructive plate margin.
- North American Plate moving west; Eurasian Plate moving east.

Primary Effects

- Volcanic ash plume at 11 000 metres.
- Erupted under the glacier, causing severe flooding.
- Damage to roads, bridges, water supplies and livestock.
- Lava flows.
- Very fine-grained ash was a hazard to air traffic.

Secondary Effects

- Air space closed in Europe and thousands of flights cancelled.
- Glaciers covered in ash for months (increasing glacial melt).

Immediate Responses

- 500 farm families evacuated.
- National Emergency Agency dredged rivers, cleared ash and installed temporary bridges.

The ash cloud from the Eyjafjallajokull eruption caused the cancellation of flights all over the world

Long-Term Responses

- Individual nations grounded air traffic according to local weather conditions following the eruption.
- A scientific review investigated the effect of ash eruptions on air traffic.
- Further research continues to find better ways of monitoring ash concentrations and improving forecast methods.

Volcano Case Study – Mount Merapi, Indonesia, Lower Income Country (LIC)

Background

- Erupted in October, 2010.
- Caused by the Indo-Australian Plate subducting beneath the Eurasian Plate.
- Destructive plate margin in the 'Pacific Ring of Fire'.

Primary Effects

- 353 people were killed.
- 360 000 people were displaced to 700 emergency shelters.
- Volcanic bombs and hot gases for up to 11 km.
- Pyroclastic flows of up to 14 km.
- Ash fell up to 30 km away.
- Villages close to the volcano were buried.

Secondary Effects

- 350 000 people left homes in the area.
- Sulphur dioxide blown across Indonesia and as far as Australia.
- Ash cloud disrupted air transport.
- Roads blocked.
- Lahars.
- Food prices increased.
- Airports closed.

Immediate Responses

- Evacuation centres were established.
- 20 km exclusion zone was set up around the volcano.
- International aid arrived from charities like the Red Cross.

Long-Term Responses

- People moved to newer and safer homes.
- Funds were provided for farmers to replace livestock.
- Data from the eruption contributed to computer models to improve predictions.
- Improved warning and evacuation procedures.
- Construction of dams to hold back future lahars.

Quick Test

1. Which has a wider base: a composite or a shield volcano?
2. What is the difference between pyroclastic flow and a lahar?

Key Point

The effects of and responses to volcanic eruptions vary between areas of contrasting wealth.

The eruption of Mount Merapi sent down tonnes of ash

Key Words

composite volcano
shield volcano
pyroclastic flow
lahar
active volcano
dormant volcano
extinct volcano

Tropical Storms

You must be able to:

- Describe and explain where tropical storms occur and why they are formed there
- Describe and explain the structure and features of a tropical storm
- Consider the possible links between climate change and tropical storms.

Global Distribution of Tropical Storms

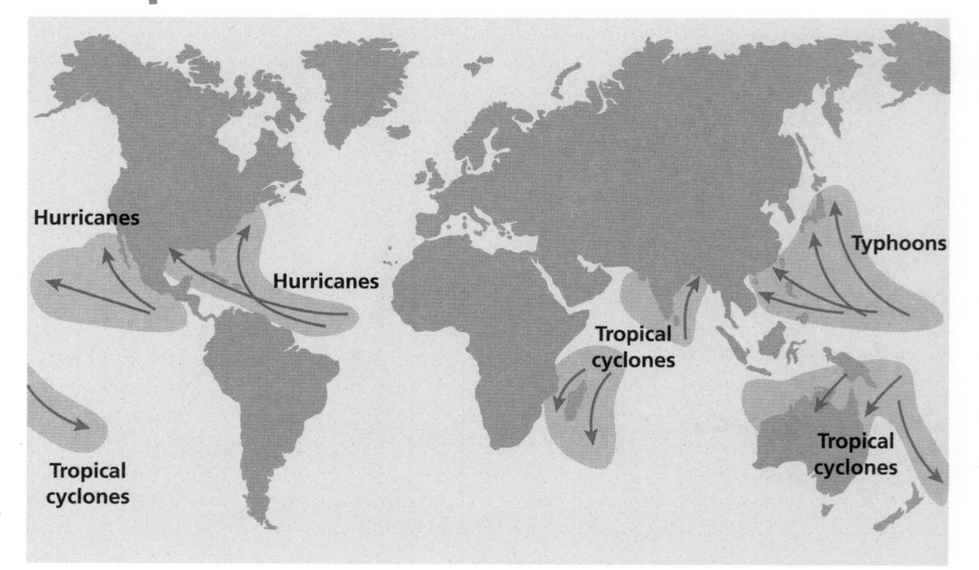

- **Atmospheric circulation** is the movement of air around the Earth in cells, transferring and redistributing energy. At sea level there are alternate low and high-pressure belts. It may be helpful to think of the atmosphere as an 'ocean' of air with waves, whirlpools and currents.
- **Tropical storms** are intense low-pressure systems, with distinct structure and features, formed over warm ocean waters in low latitudes. They are linked to the atmospheric circulation because warm, humid converging air at sea level starts to rise, creating low-pressure systems. If the average wind speed exceeds 74mph (119km/h), the storms are called 'cyclonic'.
- The Northern Hemisphere experiences more tropical storms and most occur between June and November. In the Southern Hemisphere, the tropical storms usually occur between November and April.
- Region-specific names for tropical storms are:
  - **Cyclone**: Indian Ocean and near Australia
  - **Hurricane**: Atlantic, Caribbean and Pacific (near North and South America)
  - **Typhoon**: Pacific (near Asia).

Key Point

Tropical storms are intense low-pressure systems, with a distinct structure and features, formed over warm ocean waters in low latitudes.

What Causes Tropical Storms?

- Tropical storms form over warm ocean water (≥ 26–27°C) between latitudes 5° and 20° north and south of the Equator.
- Moist air evaporates and rises, creating low pressure.
- More warm, moist air is sucked in. A 'fountain' of air is formed.
- As it rises and cools, condensation occurs, creating clouds and convectional rain.

Hurricane Katrina, August 2005: a Category 5 Tropical Storm Which Devastated New Orleans in Southern USA

Anti-clockwise rotation in Northern Hemisphere

Eye

The Structure and Features of a Tropical Storm

- A tropical storm can be around 600 miles (1000 km) in diameter and has a distinctive structure.
- Owing to the Earth's rotation, **Coriolis force** occurs, which makes low-pressure systems spin anti-clockwise in the Northern Hemisphere and clockwise in the Southern Hemisphere.
- If a tropical storm becomes 'cyclonic', it spins so fast that the air around the centre forms a vortex, which creates an eye 20–40 miles (30–65 km) wide.
- The intense low pressure creates a 'dome' of seawater and a **storm surge** occurs, which can lead to coastal flooding.
- Cooler, drier air is dragged down into the eye. Consequently, inside the **eye of the storm** is a place with strangely calm weather and very little cloud or rain, whereas around the edge of the eye the winds are strongest.
- The Pacific Ocean has more severe tropical storms than any other basin.
- When a tropical storm makes landfall, it loses its energy supply (warm water) and slows owing to friction (especially if the land is hilly).

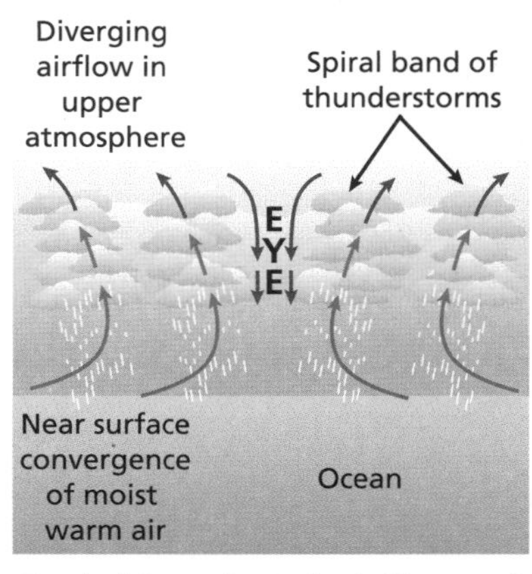

Tropical Storm Intensity is Measured Using the Saffir–Simpson Scale

Category	Wind speed
Five	≥70 m/s, ≥137 knots ≥157 mph, ≥252 km/h
Four	58–70 m/s, 113–136 knots 130–156 mph, 209–251 km/h
Three	50–58 m/s, 96–112 knots 111–129 mph, 178–208 km/h
Two	43–49 m/s, 83–95 knots 96–110 mph, 154–177 km/h
One	33–42 m/s, 64–82 knots 74–95 mph, 119–153 km/h

Climate Change and Tropical Storms

- Warm ocean water drives tropical storm formation.
- Warmer oceans expand, so storm surges may be worse.
- Climate change may alter the distribution of tropical storms and their **frequency** and **intensity** may increase, but the evidence for this is inconclusive.

Key Point

If the oceans become warmer due to climate change, this may affect the distribution, frequency and intensity of tropical storms.

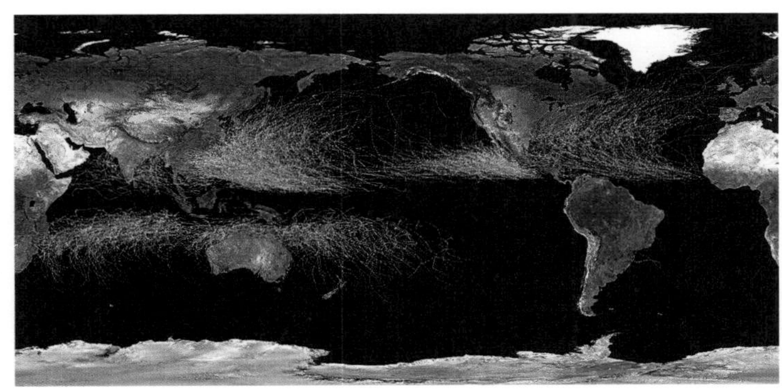

The cumulative tracks of all tropical storms for the period 1985–2005

Key Words

atmospheric circulation
tropical storm
cyclone
hurricane
typhoon
Coriolis force (Coriolis effect)
storm surge
eye of the storm
frequency
intensity

Quick Test

1. Which hemisphere has the most tropical storms?
2. Which ocean has the highest number of severe tropical storms (≥ Category 4)?
3. Why might a storm surge cause more deaths than high winds?
4. Why do tropical storms weaken on landfall?
5. What is the difference between frequency and intensity?

Tropical Storms – Case Study

You must be able to:

- Demonstrate your understanding of a case study of a tropical storm
- Describe and explain the primary and secondary effects
- Describe the immediate and long-term responses
- Evaluate some of the responses and explain how the 3 Ps can reduce the negative effects.

Case Study: Typhoon Haiyan

Context

- The Philippines in south-east Asia: an archipelago (island chain) made up of more than 7000 islands (around 4000 are inhabited)
- Population: 100 million (2015 estimate)
- Capital: Manila
- Area: 115 830 square miles or 300 000 km²
- Major religion: Catholicism (often a key player in supporting social infrastructure and resilience, especially in poorer communities)
- Average income: US $7500
- Typhoon Haiyan struck in November, 2013.

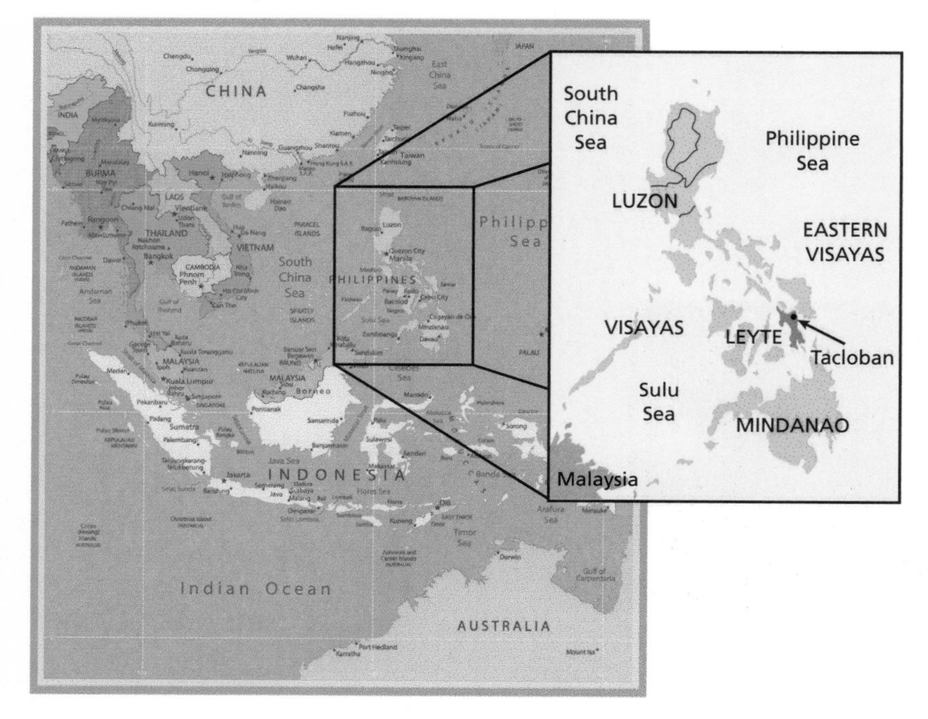

Key Point

Use the six Ws to structure your revision: Where? When? What? Who? Why? How?

The Storm and Why it Happened

- Typhoon Haiyan was a Category 5 tropical storm.
- It was the strongest storm ever recorded at landfall.
- Wind speeds were among the highest ever recorded in a tropical storm [1-minute sustained: 315 km/h (195 mph)].
- Worst storm in the Philippines for over 130 years.
- Warm surface water in the western Pacific.
- Climate change may have contributed.

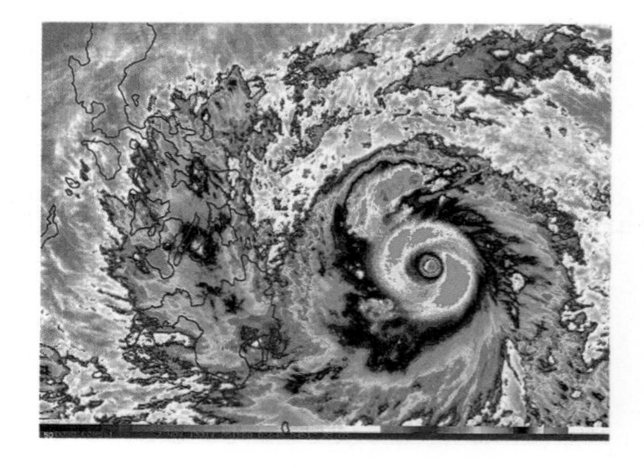

A satellite image of Typhoon Haiyan

Primary and Secondary Effects

- **Primary effects** cause widespread devastation – most settlements on the island of Leyte were destroyed.
- Landslides occurred across the landscape.
- Storm surges of 5–6 m on the islands of Leyte and Samar.
- Tacloban Airport terminal was destroyed.
- The first floor of the Tacloban City Convention Center, which was serving as an evacuation shelter, was submerged and many people were drowned.
- **Secondary effects** damage **infrastructure**, impeding relief efforts.
- Economic effects included high losses due to businesses being damaged or closed, and development was halted.
- Social effects included homelessness (1.9 million people); displacement (6 million people); bereavement; disease due to lack of food, water, shelter and medication; and schools closed.
- Ecosystems were damaged and farmland lost.
- Human factors made matters worse (**3 Ps**):
  - prediction: high-level warnings were too late
  - protection: communications infrastructure was too vulnerable (it failed in the Visayas islands)
  - planning: many people chose to stay in their homes (**inertia**).

Death Toll and Responses

- At least 6300 were killed (the exact number is unknown). The number of dead from the Eastern Visayas was 5877.
- Most of the dead were victims of the storm surge.
- Immediate responses:
  - Much of the central Philippines (Visayas) was placed under a state of national emergency.
  - Worldwide relief effort with aid of over $500 m (£400 m).
- Long-term responses: how did management strategies reduce the risk, and how can they do so in future?
  - The authorities were heavily criticised for being too reactive, rather than proactive.
  - Following a review of the 3 Ps, the authorities adopted much more proactive strategies with 'zero casualty' targets. They tackled the issue of inertia with incentives such as free bags of rice to persuade people to leave homes and property behind.
  - 'Evacuation rather than rescue, that's our doctrine,' said an emergency management chief.
  - The strategies were very successful: just over a year after Haiyan, Typhoon Hagupit hit the Philippines. It was also a Category 5 storm, but in marked contrast only 18 people were killed.

Quick Test

1. Which country was most severely hit by Typhoon Haiyan?
2. Give a measure of its magnitude.
3. Describe one primary and one secondary effect.
4. Give one of the causes.
5. Outline one immediate and one long-term response.

Devastation after Typhoon Haiyan

Key Point

The effects of Typhoon Haiyan (local name: Yolanda) on the Philippines were significantly affected by the country's **resilience**, which depends on its level of development and preparedness.

UK overseas aid workers loading emergency aid on to trucks after Typhoon Haiyan

Key Words

infrastructure
resilience
3 Ps
inertia

Extreme Weather in the UK

You must be able to:

- Demonstrate that you can explain the main influences on the UK climate
- Describe the main types of atmospheric hazards affecting the UK
- Analyse the causes, impacts and responses of an extreme weather event in the UK.

UK Climate and Weather Hazards

- The UK has a temperate climate, i.e. relatively moderate temperatures and levels of precipitation.
- A key factor is its location in the **mid-latitudes** on the coast of north-west Europe facing the Atlantic, where cold and warm **air masses** create **fronts**, which produce precipitation.
- The **prevailing winds** (most common winds) are westerlies from the Atlantic Ocean, and the relatively warm, moist air from the **North Atlantic Drift** (or **Gulf Stream**) creates a dominant **maritime** effect.
- Less frequently, the UK receives relatively dry **continental** air from land masses to the south and east. In winter, such air can be extremely cold but in summer it is very hot.
- **Altitude** is also influential. Higher ground in the west has greater precipitation (**relief rainfall**) and lower temperatures. Consequently the east is in a **rain shadow** (receiving much lower precipitation).
- The main hazards affecting the UK are atmospheric or **hydro-meteorological hazards** linked to the water cycle.
- The main types of weather hazard event include: river flooding; sea flooding; winter storms; snow and ice; **drought**.

Key Point

The main weather hazards impacting the UK include river and sea flooding, winter storms, snow, ice and drought.

The UK Drought of 2010–12

- Droughts are extended dry periods lasting for months or years.
- The 2010–12 drought was one of the ten most significant for 100 years.
- The pattern of such events may contribute to evidence that weather is becoming more extreme in the UK due to climate change.

Causes

- **Blocking highs** (slow-moving anticyclones) led to very dry winters in 2009–10 and 2011–12.
- East winds were common, bringing in dry continental air.
- A significantly lower precipitation than normal (see map, right).

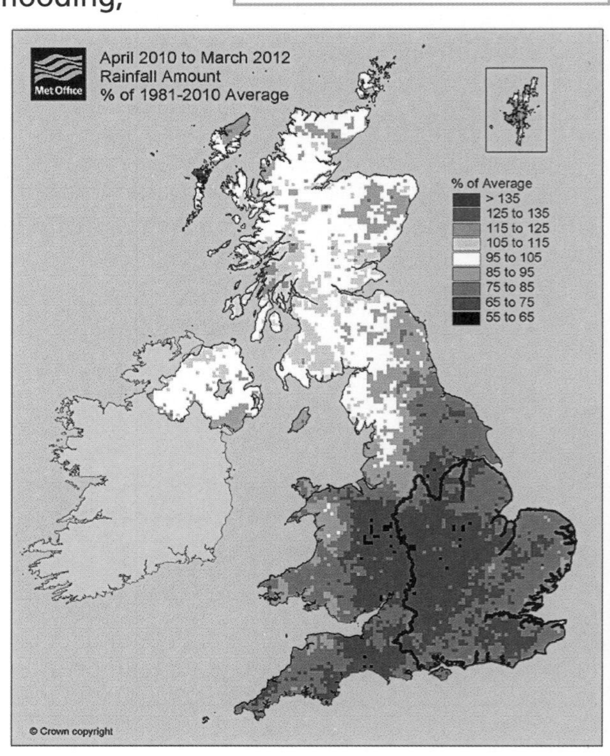

- Climate change may have contributed.
- An imbalance between the demand and supply of water in the densely populated south-east and Midlands, which suffer **water deficit** and **water stress**. These areas rely on piping of water from regions of **water surplus** and/or groundwater supplies.

Impacts

- Economic and social impacts:
  - farmers struggled to provide water for livestock and to harvest crops
  - low reservoir levels.
- Environmental impacts:
  - groundwater and river levels were very low, affecting aquatic ecosystems
  - wildfires spread.

Management Strategies

- Hosepipe bans were introduced (affecting six million consumers).
- Water meters were installed to monitor usage.
- Water companies fixed leaking pipes.
- Water-saving devices were encouraged.
- Education on water use.
- Improved waste-water recycling.
- New reservoirs and pipelines were considered.
- Desalination plants were being considered (but at high cost).

Key Point

Extreme weather events such as drought have economic, social and environmental impacts.

Quick Test

1. What is meant by a maritime influence on the UK climate?
2. a) Give an example of one extreme weather event in the UK.
   b) Describe one impact and one management strategy for the extreme weather event.

Key Words

mid-latitudes
air masses
fronts
prevailing winds
North Atlantic Drift
Gulf Stream
maritime
continental
altitude
relief rainfall
rain shadow
hydro-meteorological hazards
drought
blocking high
water deficit
water stress
water surplus

Climate Change

You must be able to:

- Describe and explain the main evidence for, and causes of, climate change
- Assess different approaches to the management of climate change.

Evidence for Climate Change

- The Intergovernmental Panel on Climate Change (**IPCC**) is an internationally accepted authority on climate change.
- The IPCC found that Northern Hemisphere temperatures during the 20th century were the highest for the past 1300 years.

Long-Term Evidence

- In the absence of reliable climate data, **proxy measures** are used, e.g. ice cores, marine sediment cores and pollen analysis.
- Proxy measures of temperature since the start of the **Quaternary period** (the last 2.6 million years) show that the Earth has experienced an 'Ice Age' of several lengthy **glacials** (extremely cold periods of glacier growth) interspersed with shorter, warmer **interglacials** (warmer periods of glacier retreat).
- The Earth is currently in an interglacial period.

Short-Term Evidence (Last Few Hundred Years)

- As climate measurements were unreliable before 1850, other proxy measures are needed such as tree rings and historical sources (e.g. landscape paintings and literature).
- Data show a marked global temperature rise since 1850 – the 'hockey stick' graph shows variations in the Northern Hemisphere temperature over the last 1000 years.
- Warming oceans, sea-level rises, reduction in Arctic sea ice, increases in the frequency and intensity of extreme weather events, are all evidence of climate change.
- The atmosphere creates a natural **greenhouse effect**, allowing for short-wave solar radiation (e.g. in light), but **greenhouse gases** prevent the escape into space of some of the heat created (long-wave infrared radiation).
- Industrialisation and population growth have increased greenhouse gas emissions.
- Carbon dioxide increase was first noticed in the 1950s and 60s by scientists in Hawaii, with record highs in 2015.
- Global temperatures in 2015 reached 1°C above the 'pre-industrial' period of 1850–1900.
- The 11 warmest years in the instrumental record have occurred since 1998.

The 'Hockey Stick' Graph Showing Variations in the Northern Hemisphere Temperature Over the Last 1000 Years

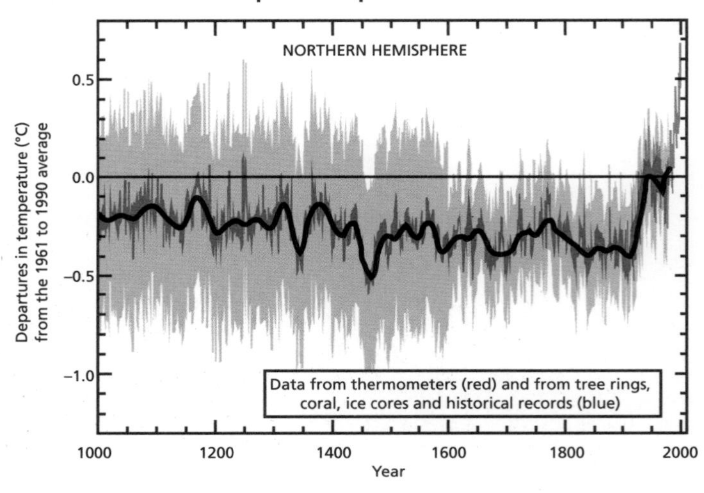

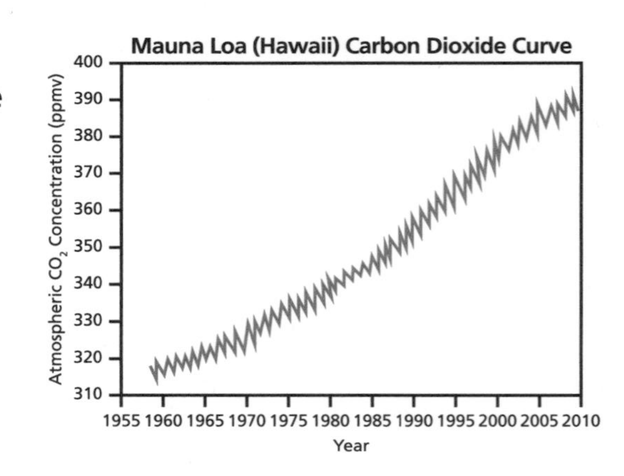

Climate Change Causes

- Natural factors include:
  - orbital changes (Milankovitch cycles)
  - volcanic activity (volcanic emissions block sunlight)
  - solar output (changes in the Sun's energy).
- Human factors (**anthropogenic** factors) include:
  - the use of fossil fuels producing greenhouse gases, such as power generation and transportation
  - agriculture, such as methane from cattle and rice paddies
  - deforestation, e.g. tree loss reduces natural **carbon capture** (**carbon sequestration**).
  - methane release from melting permafrost and ocean floors, e.g. due to anthropogenic global warming.

Key Point

Climate change is the result of natural and human factors.

Managing Climate Change Impacts

- The IPCC recommends **adaptation** and **mitigation**.
- Adaptation means making adjustments to reduce potential damage, to limit the impacts and to take advantage of new opportunities, e.g. alternative energy production, planting trees and changes in agricultural systems.
- Mitigation means cutting greenhouse gas emissions, reducing or eliminating long-term risk to human life and property, internationally agreed targets (such as the Paris Agreement ratified in November, 2016) to reduce greenhouse gas emissions, managing the water supply, reducing the risk from rising sea levels and carbon capture.
- Carbon capture removes carbon at the emission source and stores it deep underground.
- Critics of adaptation say that without mitigation it only treats the symptoms, not the causes, of climate change.
- Critics of mitigation say that reversal of climate change is not possible. A minority still deny climate change completely.

Cuts to greenhouse gas emissions are a key part of tackling climate change

Quick Test

1. What does the IPCC stand for?
2. Give two examples of proxy measures.
3. Describe and explain one physical and one human cause of climate change.
4. Which responses are **not** adaptation?
   Which responses are **not** mitigation?

Climate change response	NOT adaptation	NOT mitigation
Electric cars		
Higher sea walls		
Tidal power		
Wind farm		
IPCC carbon reduction targets		
Improving air-conditioning in houses		

Key Words

IPCC
proxy measures
Quaternary period
glacials
interglacials
greenhouse effect
greenhouse gas
anthropogenic
carbon capture / carbon sequestration
adaptation
mitigation

Practice Questions

Tectonic Hazards 1

1. What causes the Earth's plates to move? [2]
2. Draw a diagram to help explain what happens at a destructive plate margin. [4]
3. Draw a diagram to help explain what happens at a constructive plate margin. [4]
4. Draw a diagram to help explain what happens at a conservative plate margin. [4]

Total Marks .... / 14

Tectonic Hazards 2

1. What do the terms LIC and HIC mean? [2]
2. With reference to the Tohoku, Japan 2011 earthquake, which statements are true and which are false?

   A A nuclear power station was seriously damaged.
   B The epicentre was at sea.
   C It measured over 6 on the Richter scale.
   D It produced tsunami waves up to 40 metres in height.
   E The tsunami waves only affected a nuclear reactor. [5]

3. Explain why the largest earthquakes do not always result in the most deaths. [6]

Total Marks .... / 13

Tectonic Hazards 3

1. Describe the effects of the Mount Merapi volcanic eruption in 2010. [4]
2. Identify two primary effects following the eruption of Eyjafjallajokull. [2]
3. What long-term responses can be adopted to reduce the effects of a volcanic eruption? [4]

Total Marks .... / 10

Tropical Storms

1. Which of the following statements is true?

   A Tropical storms form along the Equator.
   B Tropical storms form between latitudes 5° and 20° north and south of the Equator.
   C Tropical storms form between latitudes 30° and 50° north and south of the Equator.
   D Tropical storms form between latitudes 40° and 60° north and south of the Equator.
   E Tropical storms form between latitudes 50° and 70° north and south of the Equator. 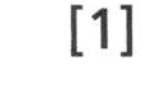 [1]

2. Give three key features of a tropical storm, such as the one below.  [3]

3. What are the region-specific names used for tropical storms in these areas?

   a) Indian Ocean and near Australia
   b) Atlantic, Caribbean and Pacific (near North and South America)
   c) Pacific (near Asia) [3]

4. How does the Coriolis force (Coriolis effect) affect wind direction around tropical storms in different parts of the world? [4]

Total Marks .... / 11

Practice Questions

Tropical Storms – Case Study

1. What are the '3 Ps'? [3]
2. For a tropical storm you have studied, state its name, where and when it occurred. [3]
3. Select the main category for each of these impacts of a tropical storm:

Impacts	Economic	Social/Political	Environmental
Homelessness			
Factories and other businesses closed or inaccessible due to damage to transport infrastructure			
Waterborne diseases			
Damage to ecosystems			
Schools closed for weeks			

[5]

4. For these effects of a tropical storm, indicate which are primary (P) and which are secondary (S).

a) People drowned in storm surge.
b) Thousands of people are homeless for over a year.
c) Outbreaks of cholera and dysentery kill hundreds of people.
d) Severe damage to infrastructure, such as railways and bridges, by flash floods.
e) Homes destroyed by high winds.
f) Schools and businesses are closed for months. [6]

Total Marks .... / 17

Extreme Weather in the UK

1. For an extreme weather event in the UK that you have studied, state when it occurred. [1]
2. How does altitude of the land influence the UK climate? [2]
3. What are the main types of weather hazard affecting the UK? [3]
4. Outline the main causes of an extreme weather event in the UK. [4]

Total Marks .... / 10

Climate Change

1. IPCC is the abbreviated name of which organisation?

   A International Policy on Climate Change
   B International Panel on Climate Change
   C Intergovernmental Policy on Climate Change
   D Intergovernmental Panel on Climate Change
   E Intermediate Panel on Climate Change [1]

2. What key term is sometimes used to describe human causes of climate change? [1]

3. Give three human causes of climate change. [3]

4. Draw lines to match the key terms to the correct definitions.

Key term	Definition
Glacial	Indirect ways to find out average temperatures from the past
Interglacial	Last 2.6 million years, including the 'Ice Age'
Quaternary	Warmer periods of glacier retreat
Proxy measure	Extremely cold periods of glacier growth

[3]

Total Marks .... / 8

Ecosystems and Balance

You must be able to:

- Describe how ecosystems are structured
- Explain, using a small-scale UK example, how ecosystems can have their balance affected through changing one component.

What is an Ecosystem?

- **Ecosystems** are communities of living and non-living components that all function together to create a distinctive environment.
- **Food chains** show simple relationships between different organisms. Basically that means, what eats what.

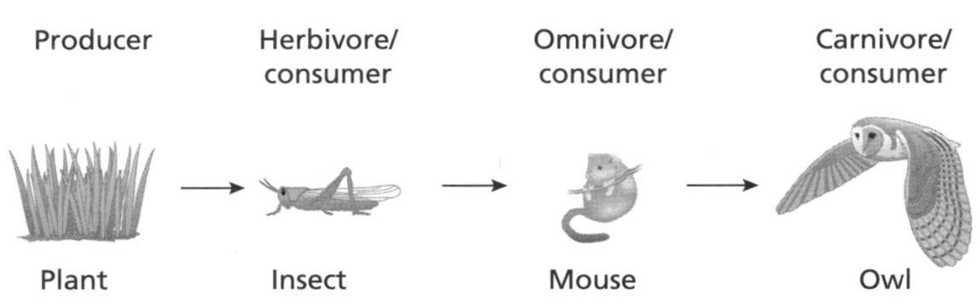

- **Food webs** show more complex **interrelationships** between organisms.

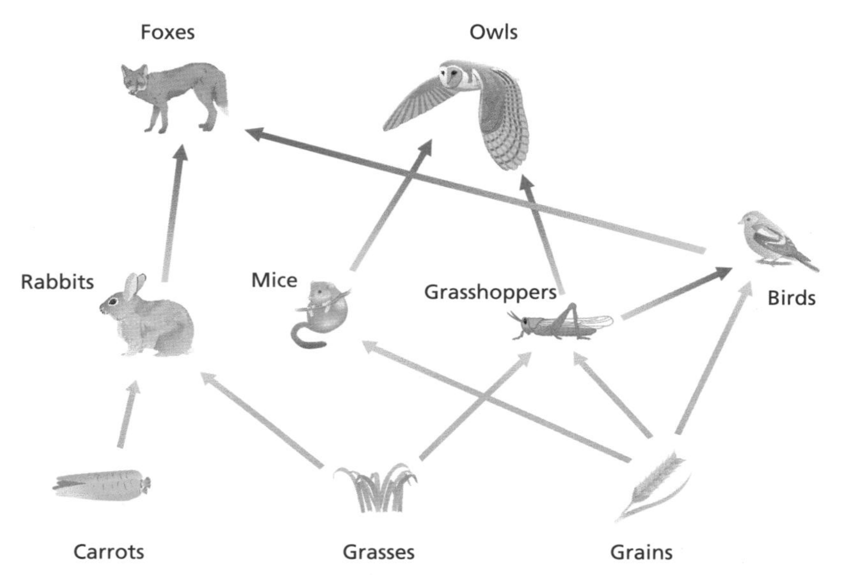

- Ecosystems can be small scale, such as a pond or hedgerow in the UK.
- Ecosystems can be large scale, such as a tropical rainforest, and these are known as 'biomes'.
- Ecosystems are balanced – each organism plays its part and can be easily disrupted, particularly by the actions of humans.

Key Point

Ecosystems are a critical part of the natural world. All species alive today form part of an ecosystem. The climate and soil form the building blocks of all ecosystems.

A pond is a small-scale ecosystem

- **Producers** are organisms such as plants that convert the Sun's energy into sugars, thus *producing* food for them.
- **Consumers** are species that eat other species. They can be **herbivores** (such as goats) that eat plants or **carnivores** (such as lions) that eat other animals.
- **Decomposers** such as fungi act to break down the remains of dead plants and animals and then return their nutrients to the ecosystem.

A hedgerow is another type of small-scale ecosystem

Balance in Ecosystems

- Maintaining balanced ecosystems is critical to the survival of life on Earth.
- A hedgerow is one small-scale example of an ecosystem in the UK.
- In a hedgerow, the plants (such as small trees) are the producers, small animals (such as mice) are herbivore consumers and birds (such as owls) are carnivore consumers.
- Ecosystems naturally manage themselves in a variety of complex ways: populations of particular organisms are kept in check by being the prey of another – too many beetles born in one year means more food for the birds that prey on them.
- Nutrients move around an ecosystem through the **nutrient cycle** – plants grow and are eaten by herbivores; the herbivores are then eaten by carnivores; the carnivores die and decompose; the decomposed remains are reabsorbed by the trees. At each stage, nutrients are transferred around the ecosystem.
- Every part of the Earth's surface is part of an ecosystem. The important ecosystems for this exam are tropical rainforests, hot deserts, polar and tundra.
- Human activity can disrupt the balance of an ecosystem. An example of this is deforestation, as the removal of trees destroys the habitats of countless organisms, thus leading to potential imbalance. The logical endpoint is the disappearance of a particular ecosystem.

Key Point

An ecosystem functions properly when all the parts of it work in harmony. This harmony can be altered by human actions.

Key Words

ecosystem
food chain
food web
interrelationship
producer
consumer
herbivore
carnivore
decomposer
nutrient cycle

Quick Test

1. Give an example of a large-scale and a small-scale ecosystem.
2. What is the difference between a food web and food chain?
3. Suggest what might happen if one species reproduces too much.
4. How do nutrients move around an ecosystem?

Ecosystems and Global Atmospheric Circulation

You must be able to:

- Describe how global atmospheric pressure changes between January and July
- Describe the locations of a range of global ecosystems.

Global Atmospheric Circulation

- Differences in **atmospheric pressure** exist. Atmospheric pressure fluctuates constantly but patterns of air movement can be observed.
- Pressure patterns can be broadly predicted and follow seasonal trends.
- Wind is the movement of air owing to many factors, such as the movement of air masses.
- Air masses can be disrupted and their behaviour distorted by features such as mountains, coasts and rivers.
- **Wind patterns** can be predictable but may be extremely erratic.
- The locations of the world's major ecosystems are heavily influenced by the existence and movement of the Earth's major high and low-pressure areas and the resulting wind patterns.

Key Point

Atmospheric circulation affects where ecosystems are found.

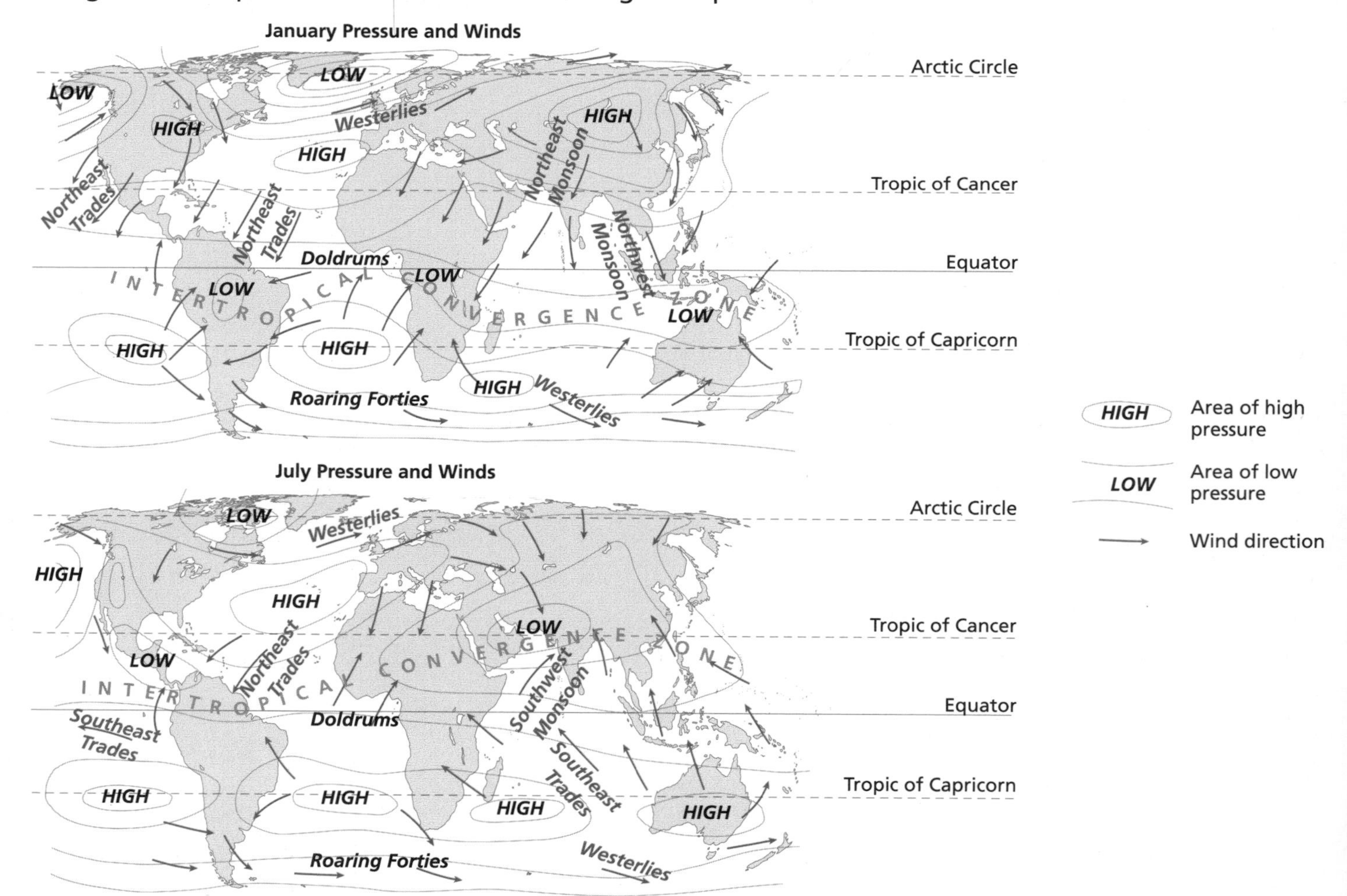

Global Ecosystems

Tropical rainforests

- Found in a wide belt encircling the Earth, roughly following the Equator and for the most part between the Tropics of Cancer and Capricorn.
- The Amazon, in the northern part of the Southern Hemisphere, is the world's largest rainforest.
- The climate is hot and wet all year round.

Tropical grasslands

- Named after the Tropics of Cancer and Capricorn, along both of which they can be found.
- The most famous tropical grasslands are those of the savannas of central Africa.
- They have long dry and brief wet seasons.
- They have little tree cover.

Temperate grasslands

- Found in two wide belts to the north of the Tropic of Cancer and to the south of the Tropic of Capricorn.
- They flourish in the centres of the continents of North America (prairies) and Asia (steppe).
- They have warm summers with very cold winters.
- They have little tree cover.

Temperate forests

- Found mostly in the Northern Hemisphere, well to the north of the Tropic of Cancer but to the south of the Arctic Circle.
- Examples of note are the deciduous forests of north-western Europe.
- They have warm, damp summers and mild winters.
- They have deciduous trees.

Boreal forests

- Found in a thin belt just to the south of the Arctic Circle, fringing tundra-type ecosystems.
- Examples of places with boreal forests include central Canada.
- They have mild summers and very cold winters.
- The trees are mostly coniferous.

Tundra

- Ecosystems found in a wide belt encircling the Earth in what are known as the high latitudes – areas on and a little south of the Arctic Circle.
- Much of northern Russia, Canada and Iceland have tundra-type environments.
- They are very cold regions.
- They have little tree cover.

Polar regions

- Found in the far north, above the Arctic Circle, and far south, below the Antarctic Circle.
- Greenland in the north and Antarctica in the south are examples of polar regions.
- The climate is very cold all year round, but with little precipitation.

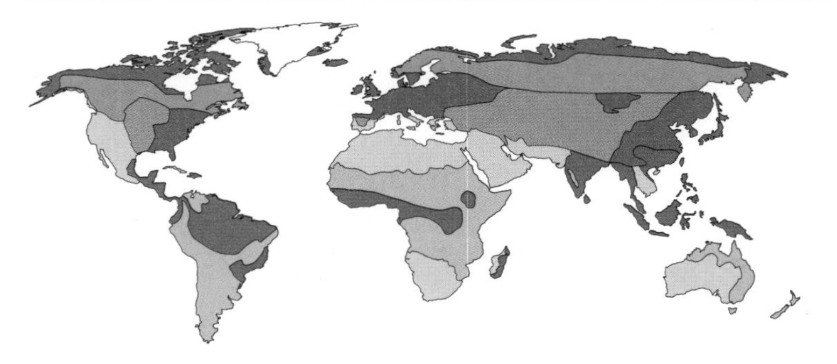

- Tropical rainforest
- Tropical grassland
- Hot desert (see pages 30-31)
- Temperate grassland
- Temperate forest
- Boreal forest
- Tundra
- Polar regions

Quick Test

1. Where are tropical rainforests found?
2. What things can affect the behaviour of air masses?
3. Give examples of places that have temperate forests.

Key Words

atmospheric pressure
wind patterns

Rainforests and Hot Deserts – Characteristics and Adaptations

You must be able to:

- Describe the distinctive characteristics of rainforests and hot deserts
- Describe how plants and animals adapt to named rainforest and desert ecosystems.

Characteristics of Tropical Rainforests

- The climate in tropical rainforests is generally warm all year round, with average daily temperatures of around 27–29°C. Rainfall is also high at around 500–600 mm each month.
- The soil layer is known as a **latosol**. Latosols are deep but have few nutrients within them. The vast majority of the nutrients are found in the **leaf litter**, which decays rapidly due to the warm, damp climate.

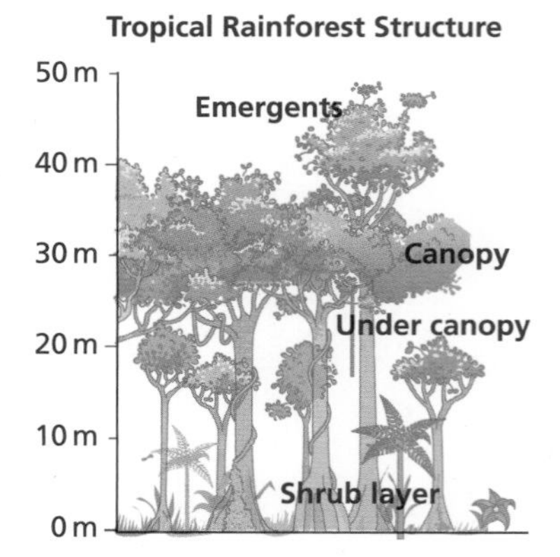

Adaptations in the Rainforest

- Biodiversity (the range of species) is very high in rainforests but is at risk from human impact.
- Trees such as the mahogany and kapok in the Amazon have developed huge **buttress roots** that provide stability but are also largely above ground so as to absorb nutrients directly from the fast-decaying leaf litter.
- Pitcher plants have adapted to the low nutrient soils by developing a taste for insects – they are attracted using scent glands and then caught using a slippery flower before being digested.
- Forest elephants eat clay from ponds in rainforest clearings to counteract the toxins in the leaves they eat.
- Ecosystems are balanced and display interdependence – animals, plants, climate and water all play their part and can be disrupted by the actions of humans.

Buttress roots

Characteristics of Hot Deserts

- The temperature in deserts is high all year round with summer temperatures reaching as high as 40°C and winter temperatures reaching up to 20°C.
- Monthly rainfall is low with an average of approximately 3 mm.
- Species have developed ways of dealing with the conditions.
- Rivers do flow through hot deserts – such as the Colorado in south-western USA – and humans in richer countries have sought to use them to help farm the land through **irrigation**.
- Rivers appear and disappear quickly and often as the result of high rainfall in mountain sources.

Key Point

Tropical rainforests are found along the Equator and between the Tropics of Cancer and Capricorn. The soils, plants and animals in a tropical rainforest have all developed to work together in harmony.

- Water in some deserts appears as snow, which some animals like the Bactrian camel have adapted to by eating the snow.
- The soil is poor and supports little vegetation.
- Over-irrigation can cause **salinisation**.
- The ecosystem is balanced – animals, plants, climate and *especially* water are all critical and easily disrupted by humans.

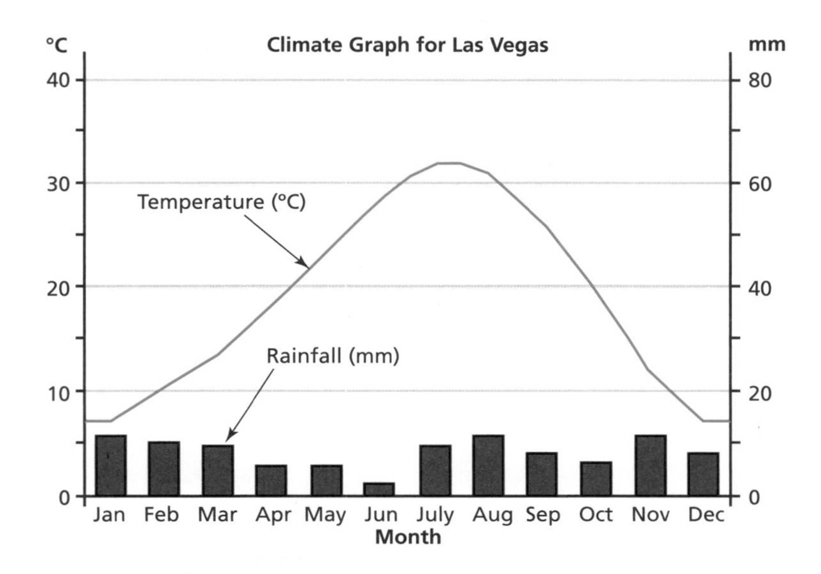

Fennec foxes

Adaptations in Hot Deserts

- The saguaro cactus of the Mojave Desert in North America has no leaves in order to cut down moisture loss through **transpiration**. It also has shallow but wide-ranging roots so that it can take advantage of brief rainstorms.
- The quiver tree of southern Africa has fleshy leaves that allow it to retain moisture.
- Fennec foxes are nocturnal, which allows them to avoid the **extreme temperatures** of the day. The fennec also has large ears, which act like a radiator in a car – letting heat escape.
- Biodiversity is low due to extreme temperatures and limited water.

Key Point

Deserts are found in different parts of the world for different reasons but they all share the same characteristics of extremely low rainfall and high summer daytime temperatures. Deserts have a low biodiversity.

Key Words

latosol
leaf litter
buttress roots
irrigation
salinisation
transpiration
extreme temperatures

Quick Test

1. Describe the climate in a tropical rainforest.
2. Name two rainforest plants and their adaptations.
3. Describe the climate in hot deserts.
4. Name two hot desert plants and their adaptations.
5. Name two examples of hot deserts.

Opportunities, Threats and Management Strategies in the Amazon

You must be able to:

- Using case study evidence, describe opportunities, threats and management strategies in tropical rainforests such as the Amazon in South America.

Opportunities and Threats

- Tropical rainforests provide goods and services to humans:
  - **Hunter-gatherers** collect edible plants and catch wild animals to eat
  - Soils can be fertilised through small-scale **shifting cultivation** of crops such as manioc and cassava
  - Trees provide fuel and building materials
  - Medicines and hunting poisons can be extracted from a wide variety of plants and animals.
- Trees absorb atmospheric carbon and release oxygen.
- **Commercial** and **subsistence farming** of crops (such as soybeans) and animals (such as cows) is one of the major reasons for the deforestation of the Amazon.
- There are large deposits of minerals such as iron, bauxite, nickel and copper beneath the forest but to reach them means large-scale removal of the vegetation.
- Building roads such as the Trans-Amazonian Highway has led to increased **settlement** as Brazil's population has grown. Roads also allow access to harder-to-reach areas, thus creating new centres of expansion.
- **Hydroelectric power (HEP)** schemes such as that at Tucuruí in Brazil create huge amounts of energy for a growing economy but also flood huge areas of forest. The dams also encourage further deforestation for farming and settlement.
- Removing the vegetation causes soil erosion as roots that bound the soil together no longer do so, leaving it exposed and weakened.
- Deforestation also removes a valuable source of humus for the soil, leaving it unproductive for farming.
- With deforestation comes a loss of habitats and a resultant drop in biodiversity as species simply die out.
- The fastest way to clear forest is through burning it, but this has led to huge amounts of carbon dioxide being released into the atmosphere, thus contributing to global climate change.
- Brazil has experienced positive **economic development** as dams have provided energy for industry and homes. Large-scale mining and farming have provided jobs to millions and given Brazil a source of export income.

Deforestation

Key Point

Despite international outcry, the Amazon in South America continues to be deforested at an alarming rate.

Sustainable Management

- Rainforests such as the Amazon can be sustainably managed through some of the following measures:
  - Selective cutting (logging), whereby only trees above a certain height are felled, thus leaving smaller trees to attain maturity.
  - International agreements that name and make illegal the export of endangered plant and animal species. For example, the 1973 'Convention on International Trade in Endangered Species of Wild Fauna and Flora', better known as CITES.
  - Encouraging **ecotourism**, which makes the forest itself the tourist attraction. Tourists are shown and educated on the wonder and diversity of the ecosystem. Local populations can gain employment from such schemes, which in turn enriches the local economy without destructive exploitation and discourages **out-migration**.
  - Using labelling schemes such as the international Forest Stewardship Council (FSC), where trees grown in sustainable forests are marked as such, thus encouraging their purchase over trees felled either unsustainably or illegally.
  - International debt reduction schemes for poorer countries in order that endangered trees and animals are not seen as an easy source of export earnings, to help pay off debt.

Ecotourism can make an attraction of rainforests.

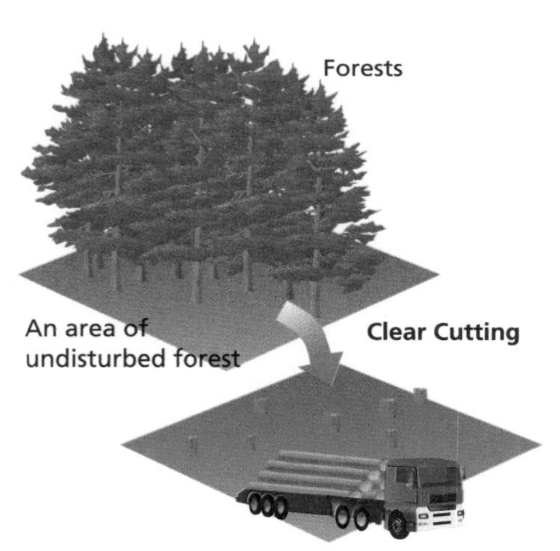

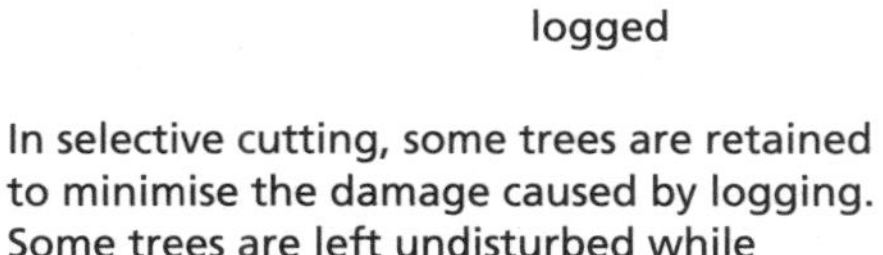

In clear cutting, a large area of the forest is cleared, harming the forest. The forest is systematically cleared of all vegetation.

In selective cutting, some trees are retained to minimise the damage caused by logging. Some trees are left undisturbed while commercially valuable ones are logged.

Quick Test

1. Give three reasons why deforestation has taken place in the Amazon.
2. Describe the impacts of deforestation.
3. How can humans use tropical rainforests without destroying them?

Key Words

hunter-gatherers
shifting cultivation
commercial farming
subsistence farming
settlement
hydroelectric power (HEP)
economic development
ecotourism
out-migration

Opportunities, Threats and Management Strategies in Hot Deserts

You must be able to:

- Using case study evidence, describe opportunities, threats and management strategies in hot deserts such as the Mojave in south-western USA or Almería in southern Spain.

Opportunities and Threats in the Mojave, South-Western USA

- In richer countries, governments build roads and bridges (**infrastructure**) that encourage new communities.
- Extensive **commercial farming** of crops such as peanuts has been made profitable by the building of the Hoover Dam (below) in Nevada, USA.
- The dam both controlled the Colorado River and created Lake Mead, which today is the source of water used for **irrigation**.
- The Hoover Dam also exists as a source of hydroelectric power (HEP).

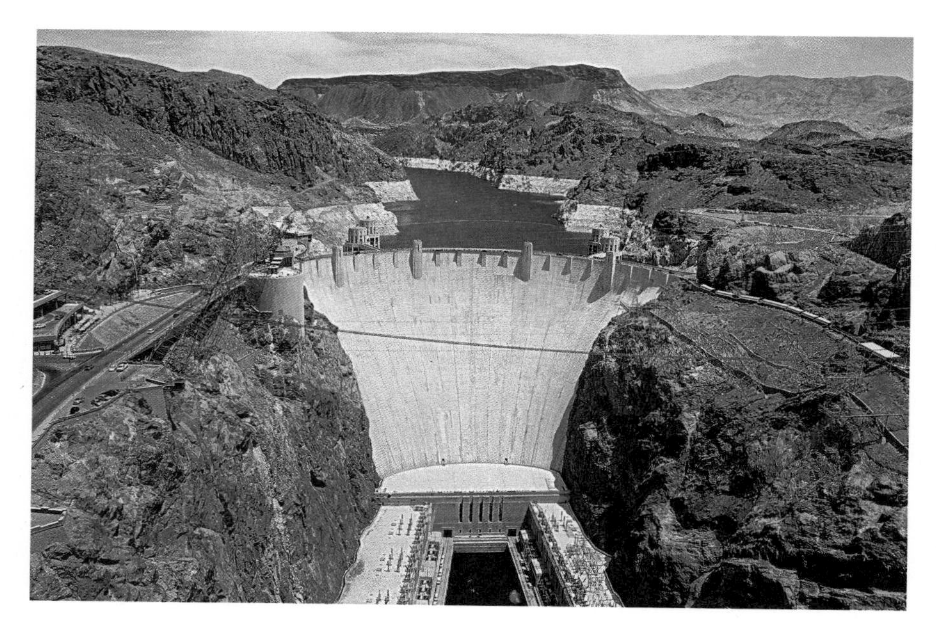

- Extreme temperatures and inaccessibility pose huge challenges to development in the Mojave.
- Tourism has become the biggest industry in the Mojave, for example the casinos in Las Vegas.
- Increased **desertification** is a real problem in places like the Mojave and other semi-arid places such as southern Spain.
- Increased water use by farming, tourism and settlement means less for the environment.

Las Vegas

Key Point

In the deserts of south-western USA, large-scale settlement has been made possible by the development of huge water projects such as the Hoover Dam on the Colorado River. By using dams to overcome the challenges of the environment, many new challenges have been created.

- Desertification has been accelerated across the globe by climate change, population growth, overgrazing by farmed cattle and over-cultivation.
- Strategies to deal with desertification include:
  - encouraging less water use among local people
  - developing planning laws that restrict the size of buildings
  - planting trees to stabilise sand dunes
  - encouraging the use of drip irrigation in farming.

Desertification in Almería, Spain

- Global warming has led to an increase in temperature and lower rainfall in Almería, southern Spain.
- **Pastoral** sheep farming has led to the removal of vegetation cover through overgrazing and trampling of the soil.
- **Terraced fields** created useful flat land for cultivation and rainfall by being built near natural watercourses.
- Many farms have been abandoned because of migration to the cities.
- Left untended, the terraces collapse, leaving the soil exposed to the Sun.
- Mining of gypsum – a mineral important to the building industry – has led to a reduction of underground **aquifers**.
- **Mass tourism** has increased the need for water. Water is needed for swimming pools, showers and for maintaining golf courses.
- Over-cultivation can lead to the process of salinisation.
- Salinisation occurs when the water in soils evaporates in high temperatures, drawing salts to the surface which are toxic to many plants.
- Desertification can be reduced through strategies such as:
  - encouraging the use of drip irrigation
  - applying **appropriate technologies** that are cheap and easy for local farmers, such as applying a coating of almond shells to the soil
  - recycling water within tourist areas
  - controlling the size and number of golf courses.

Desert Tabernas, Almería, Spain

Quick Test

1. Give three reasons why desertification has increased.
2. How does salinisation occur?
3. How can humans deal with the threat of desertification?

View over Cortijo Grande Golf Course, Almería, Spain

Key Point

Due to climate change, overgrazing by semi-nomadic pastoral farmers, the abandonment of well-tended terraced fields, large-scale mining and the growth of tourism, southern Spain is in danger of desertification.

Key Words

infrastructure
commercial farming
desertification
pastoral farming
terraced fields
aquifer
mass tourism
appropriate technology

Polar and Tundra Environments

You must be able to:
- Describe the physical characteristics of – and the adaptation of – species to polar and tundra environments and development opportunities therein
- Describe the challenges of developing a cold environment such as Antarctica and strategies to protect it, while balancing the needs of development.

Physical Characteristics of Polar and Tundra Environments

- Polar and tundra environments are both found in what are known as the high latitudes. Tundra is found about 60–70° north.

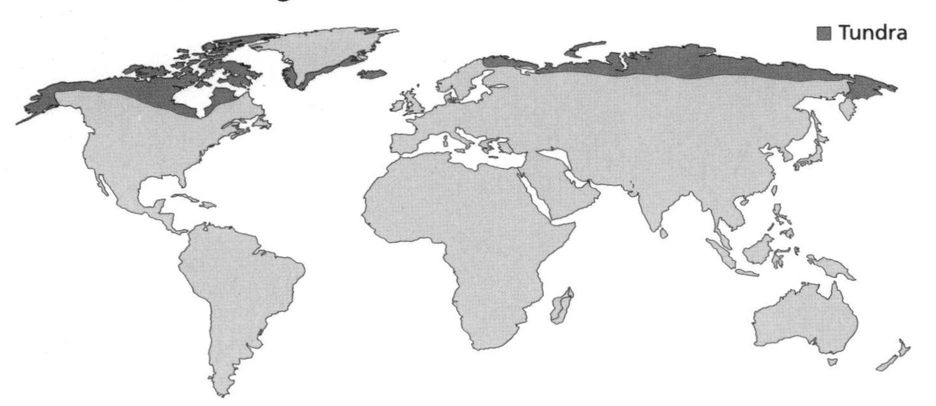

- Climates in polar and tundra environments vary throughout the year, with places like Alaska and Norway receiving moderate summers that can present daytime temperatures of 20–25°C. Winter temperatures can plunge below –40°C.
- The characteristic landscape in tundra regions is that of permafrost.

Key Point

Polar and tundra are both regarded as extreme environments. Both are cold and have characteristic ecosystems and climates. There are development opportunities present in both of these cold environments.

Adaptations

- Soil in tundra regions may remain frozen all year round. Frost-resistant plants such as the Arctic poppy survive by developing adaptations such as shallow roots and flowers that track the path of the Sun.
- Biodiversity in tundra regions is very low due to extreme cold.
- Permafrost, soil, plants and animals are all interdependent and at risk of change due to human actions.
- Animals such as the Arctic fox develop thick coats to protect against the cold.
- The Arctic hare has small ears to reduce heat loss and white fur to avoid the gaze of predators.
- There are huge deposits of minerals such as bauxite, tar sand and standard crude oil.
- Less profitable sources of oil, such as tar sand, have begun to be developed.

Arctic poppies

- The extraction of tar sand oil creates huge numbers of jobs but it destroys habitats.
- Polar ocean regions have rich seas, brimming with fish, and this attracts large-scale industrial fishing. An example is the Bering Sea near Alaska.

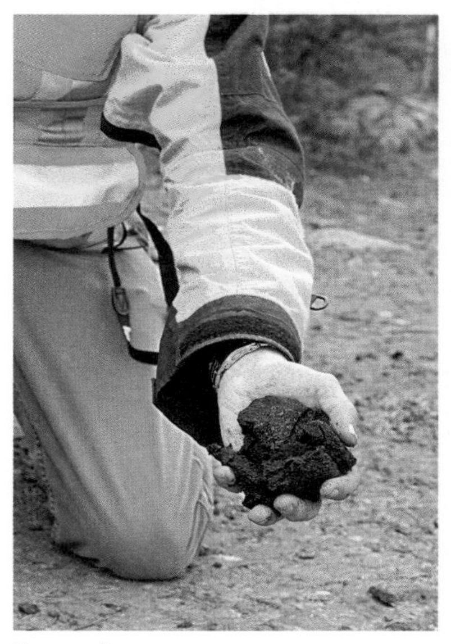
Tar sand

Antarctica – Development and Conservation

- Antarctica is the Earth's most southerly continent. There are no permanent settlements, although scientists from many countries occupy temporary stations in various places.
- Antarctica is classed as a desert due to its spectacularly low level of precipitation.
- Antarctica has huge mineral deposits such as coal, bauxite and crude oil.
- Antarctica is highly inaccessible as there is no infrastructure, no means of growing crops to sustain a population and any buildings would have to be built to withstand the extreme climate.
- Commercial fishing is big business in Antarctica, with crews from around the world (but especially Argentina and Chile), making large profits on huge catches. Governments must consider fishing quotas in order to balance the needs of fishermen and the protection of biodiversity.
- **Extreme tourism** is a recent development because tourists are being increasingly attracted to new and often hostile places.
- International pressure groups, such as Greenpeace, campaign for the protection of Antarctica.
- The 1959 Antarctic Treaty is an agreement by 12 countries declaring Antarctica a scientific preserve.
- The Madrid Protocol prohibits all mining in Antarctica to preserve it as a wilderness area.

Key Point

Antarctica is the world's last true **wilderness** and, as such, is in need of protection. Development opportunities exist but must be carefully weighed against the need to conserve the fragile habitats and ecosystems found within Antarctica.

Quick Test

1. Describe the climate of polar and tundra regions.
2. What is permafrost?
3. What are the benefits and drawbacks of extracting oil from tar sands?
4. What is extreme tourism?

Key Words

tundra
permafrost
wilderness
extreme tourism

Review Questions

Tectonic Hazards 1

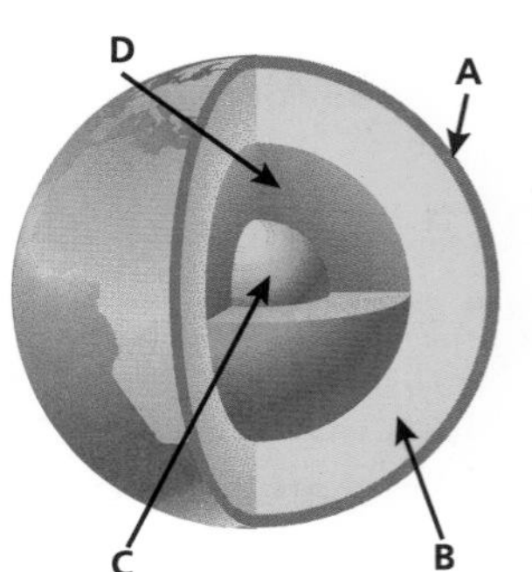

1. What is a 'hot spot'? [2]
2. State three reasons why tectonic hazards have a greater impact in LICs. [3]
3. Identify the different layers of the Earth labelled in this diagram. [4]
4. Describe how the '3 Ps' (predict, protect and prepare) can help to reduce the effects of an earthquake. [6]

Total Marks .... / 15

Tectonic Hazards 2

1. What name is given to the standard scale used to measure energy release in an earthquake? [1]
2. In what ways may the effects of an earthquake in a higher income country be different to those experienced in a lower income country? [6]

Total Marks .... / 7

Tectonic Hazards 3

1. Draw a diagram to help describe and explain the shape of a shield volcano. [4]
2. Match up the following key words with the correct definitions.

Key term	Definition
Ash cloud	Torrent of hot ash, rock, and gases and steam
Lahar	Blocks out the Sun, causing suffocation and health problems
Pyroclastic flow	Volcano nobody expects to erupt ever again
Extinct volcano	Volcano that has erupted in past 2000 years but is not currently active
Dormant volcano	Mudslide including rock debris and water

[4]

Total Marks .... / 8

Tropical Storms

1. How fast does the wind speed in a tropical storm need to be for it to be called 'cyclonic'? [1]
2. Explain why coastal flooding may be caused by tropical storms. [3]

3 Explain how climate change may influence the occurrence of tropical storms. [3]

4 Complete the table, using these terms to fill the gaps:

Cyclones **Typhoons**
Mexico **Australia**

Country	Name used for tropical storms
	Hurricanes
Philippines	
	Cyclones
Bangladesh	

[2]

Total Marks .... / 9

Tropical Storms – Case Study

1 Which of the following is most likely to have caused Typhoon Haiyan?

A High pressure in the western Pacific
B Warm surface water in the western Pacific
C Cold surface water in the western Atlantic
D Warm surface water in the western Atlantic
E High pressure in the western Atlantic [1]

2 In the context of hazard management, what is 'inertia'? [2]

3 How can human factors contribute to a higher death toll when tropical storms hit poorer countries such as the Philippines? [3]

4 Draw lines to match the following responses to their types for a tropical storm in a country such as the Philippines:

Response	**Type of response**
Foreign investment in new infrastructure, e.g. more resilient bridges and railways	Immediate international response
Government declares state of emergency across the whole country	Long-term international response
Worldwide relief effort: aid valued at over $500 million	Long-term national response
Inertia tackled by offering incentives such as free bags of rice to encourage people to leave their homes	Immediate national response

[3]

Total Marks .... / 9

Review Questions

Extreme Weather in the UK

1. Describe two of the main impacts of the UK drought of 2010–12. [2]
2. Give three strategies that were used to manage the UK drought of 2010–12. [3]
3. Explain how prevailing wind influences the UK climate. [3]
4. Draw arrows on the map to match the data below with each of the four climate zones of the UK.

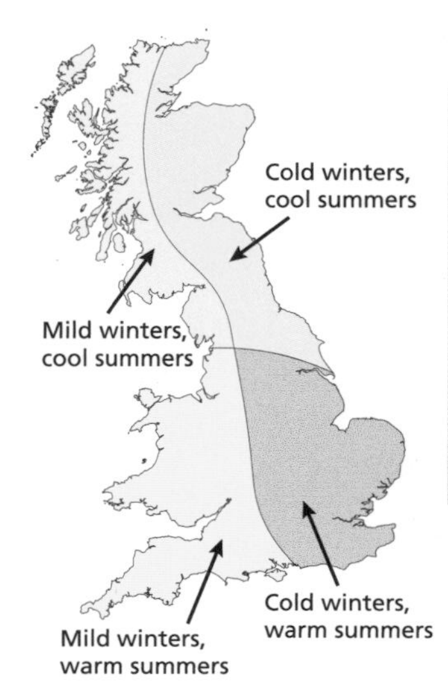

Place A	Annual rainfall 3000 mm	January average temperature 8°C	July average temperature 19°C
Place B	Annual rainfall 1000 mm	January average temperature 9°C	July average temperature 20°C
Place C	Annual rainfall 600 mm	January average temperature 3°C	July average temperature 15°C
Place D	Annual rainfall 660 mm	January average temperature 7°C	July average temperature 21°C

[3]

Total Marks .... / 11

Climate Change

1. Which of the following is the correct definition of mitigation?
   - **A** Making adjustments to reduce potential damage; limiting the impacts; taking advantage of new opportunities.
   - **B** Cutting greenhouse gas emissions; reducing or eliminating long-term risk to human life and property; internationally agreed targets to reduce greenhouse gas emissions. [1]
2. What name is given to the changes in the Earth's orbit that can cause climate change? [1]
3. How can volcanic eruptions cause climate change? [2]
4. For these management of climate change strategies, decide which are adaptation (A) and which are mitigation (M).

Developing more drought-resistant crops or irrigation schemes.	Greater use of renewable resources.
Making buildings more energy efficient.	Carbon capture and storage.
Higher flood defences along coasts and rivers.	Planting more trees.

[6]

Total Marks .... / 10

Ecosystems and Balance

1. Give an example of one small-scale ecosystem. [1]
2. Define the term 'ecosystem'. [2]
3. What is the difference between a food chain and a food web? [2]
4. Describe how the nutrient cycle works. [4]

Total Marks .... / 9

Ecosystems and Global Atmospheric Circulation

1. In January, is high or low pressure found over central Asia? [1]
2. Define 'wind'. [2]
3. Name two features that can distort the behaviour of an air mass. [2]
4. Using examples, describe the global distribution of tropical rainforests. [4]

Total Marks .... / 9

Practice Questions

Rainforests and Hot Deserts – Characteristics and Adaptations

1. What is the name of the soil found in a tropical rainforest? [1]
2. How do buttress roots help rainforest trees? [2]
3. Describe the climate in a tropical rainforest. [2]
4. Using examples, describe how animals have adapted to life in hot deserts. [3]

Total Marks .... / 8

Opportunities, Threats and Management Strategies in the Amazon

1. Name two mineral deposits found under the Amazon rainforest. [2]
2. What is selective logging? [2]
3. Describe how tropical rainforests provide goods and services to humans. [4]
4. How can hydroelectric power schemes, like the Tucuruí project, create economic development? [2]

Total Marks .... / 10

Opportunities, Threats and Management Strategies in Hot Deserts

1. What word describes roads, bridges, water and power lines? [1]
2. What do the letters HEP stand for? [1]
3. How has pastoral farming in Spain accelerated desertification? [2]
4. Describe how the problem of desertification can be tackled in the Mojave Desert. [4]

Total Marks .... / 8

Polar and Tundra Environments

1. Where are polar and tundra environments found? [1]
2. Name one benefit and one drawback of tar sand oil extraction. [2]
3. Describe the temperatures over the year in polar and tundra environments. [2]
4. Using examples, describe how plants have adapted to life in polar and tundra climates. [3]

Total Marks .... / 8

Glaciation 1: Processes

You must be able to:

- Describe the distribution of UK upland and lowland areas
- Name and locate major UK river basins
- Define and explain freeze-thaw weathering
- Define glacial erosion, transportation and deposition
- Describe and explain abrasion and plucking (quarrying)
- Name and describe material transported by glaciers
- Explain why and how material is deposited by glaciers.

UK Landscape and Glaciation

- Ice sheets covered most of the UK for some of the last ice age.
- Glaciers were active in mountain areas, even when ice sheets were smaller.

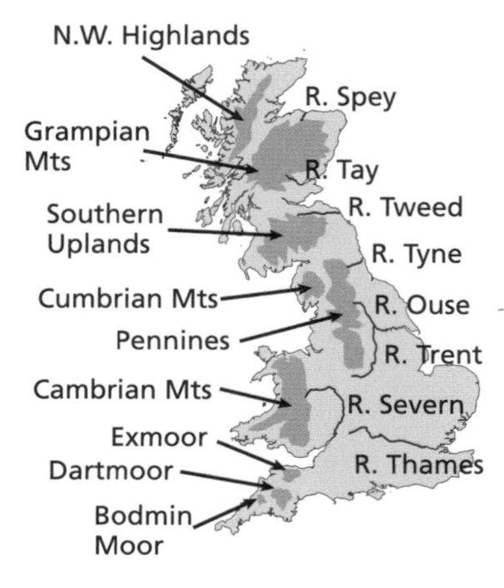

Freeze-Thaw Action

- Freeze-thaw is a weathering process taking place above the ice.
- Temperatures need to go both above and below 0°C.
- Water needs to be present.
- Rocks need to have cracks or other weaknesses.
- The more enclosed the space in which water is trapped, the more effective the process.
- Rock is shattered into angular fragments.

Glacial Erosion

- Glacial erosion is the wearing away of the landscape, which can only happen when the ice is moving.
- Ice is very effective at altering the landscape, especially if it is laden with debris. The thicker the ice, the more erosion can occur.
- Erosion is evident in upland and lowland areas.

Abrasion

- Rock fragments at the sides and base of a glacier cannot easily be pressed into the ice. If in contact with rock, such fragments will scratch the surface of the solid rock – this is abrasion. However, the debris needs to be harder than the rock being passed over.
- Scratches are created on the solid rock.

Key Point

Most upland in the UK is in the north and west, on more resistant rock.

Frost-shattered rocks (scree) on Great Gable in the Lake District

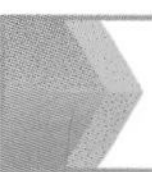

Key Point

Debris at the base of ice needs to be constantly renewed for erosion to be efficient.

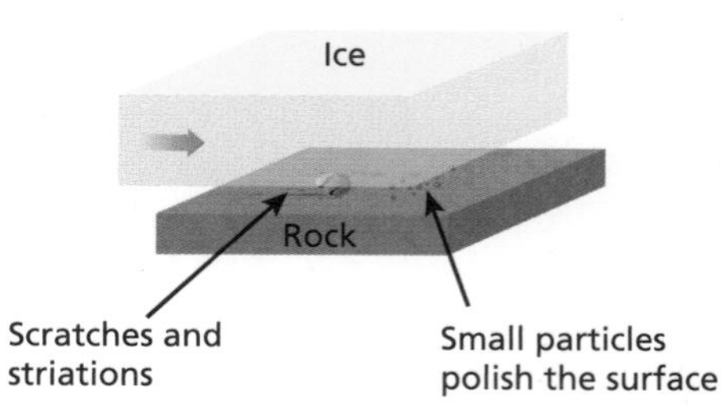

Plucking (Quarrying)

- For plucking to occur, the rock needs to have been weakened previously, perhaps by weathering.
- Moving ice melts when it meets an obstacle and the meltwater flows into cracks in the rock. As the glacier moves, the meltwater refreezes, taking the rock fragment with it.
- Ice is not strong enough to 'rip' solid rock from the landscape.
- Larger fragments can be removed by plucking than abrasion.

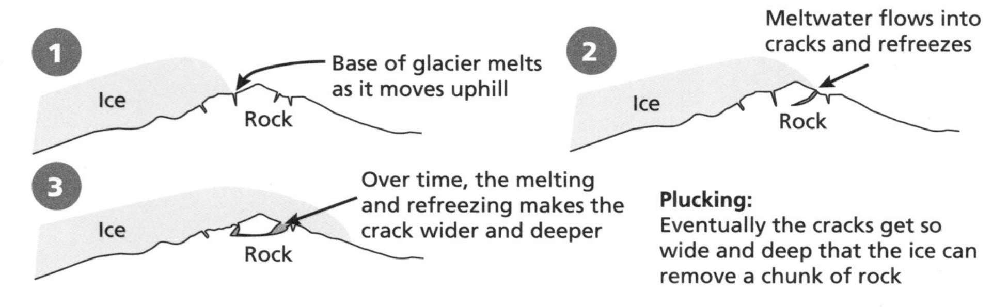

Rock surface on Snowdon showing evidence of abrasion and plucking

Glacial Transportation and Deposition

- Debris transported by a glacier is called moraine and it can vary in size from tiny particles to huge boulders.
- The debris location gives the identity of the moraine:
  - in front of the ice: terminal moraine
  - under the ice: sub-glacial moraine
  - inside the ice: en-glacial moraine
  - on top of the ice: supra-glacial moraine
  - at the side of the ice: lateral moraine
  - where two glaciers meet, their lateral moraines may form a medial moraine.
- Material moves up and down within the ice.
- Deposition results from a reduction in the size and energy of a glacier, happening most rapidly at the front and edges of the glacier. Deposition is most evident in lowland areas.
- Deposits may be in a particular form or just a blanket of material called till.
- Deposits can be carried away and redeposited as **outwash** features.

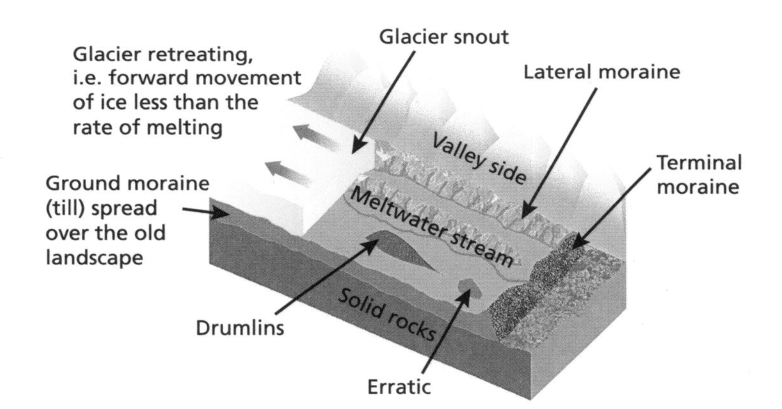

Key Point

Erosion and deposition can occur at the same time.

Quick Test

1. Why is freeze-thaw action not erosion?
2. Which erosion process removes the biggest particles: abrasion or plucking?
3. Which is more likely to show evidence of both erosion and deposition: highland or lowland?
4. How would you recognise a piece of rock that resulted from freeze-thaw action?

Key Words

glacier	moraine
freeze-thaw	terminal
weathering	sub-glacial
angular	en-glacial
erosion	supra-glacial
debris	lateral
abrasion	medial
plucking	deposition
meltwater	till
transportation	

Glaciation 2: Landscape

You must be able to:

- Recognise and describe features of glacial erosion, including glacial troughs (U-shaped valleys) and associated features: truncated spurs, hanging valleys, corries, arêtes and pyramidal peaks
- Explain the role of weathering and erosion processes in their formation
- Recognise and describe features of glacial deposition and explain distinctive forms
- Recognise the spatial and formation relationships between landforms.

U-Shaped Valleys and Troughs

- **Glacial troughs** (**U-shaped valleys**) are flat-floored and steep-sided with sudden closure at the upstream end (**trough end**) and are usually in former river valleys.
- Ice fills the valley, not just the river channel, to great thickness.
- Ice moving downslope erodes the base and sides of the valley, straightening its course. This is called 'all round erosion'.
- Freeze-thaw action above the ice brings debris on to the glacier surface, widening the valley.
- Abrasion and plucking occur where the ice touches the rock, scouring away material, including former interlocking spurs, to create **truncated spurs**.
- After the disappearance of the ice, the original valley floor is rocky and irregular. Infilling by material washed from glaciers and rivers creates a smooth, flat floor.
- **Ribbon lakes** form on the valley floor and are gradually filled in. Today, some but not all valleys have lakes.

Hanging Valleys

- Tributary valleys filled with smaller, less-powerful glaciers so were much less eroded. Their ice contributed to the main glacier. Former confluences are eroded away, leaving tributary valleys '**hanging**'.
- Ice from tributaries creates ice-falls into the main valley, which may become waterfalls.

Corries, Arêtes and Pyramidal Peaks

- High above the valleys, steep-sided, semi-circular rock basins form, called **corries**.
- Freeze-thaw action above the ice creates steep, high, back walls. Side walls are usually lower and less steep.
- At the base of the ice, abrasion deepens the basin and at the back of the ice, plucking makes the headwall steeper.
- Ice rotates through the basin to move downstream, causing over-deepening.
- Back-to-back corries make a knife-edged rock ridge (an **arête**).
- On large mountains, arêtes on three or more sides create a **pyramidal peak**.

Key Point

All erosion landforms result from both weathering and erosion.

Formation of Glacial Troughs

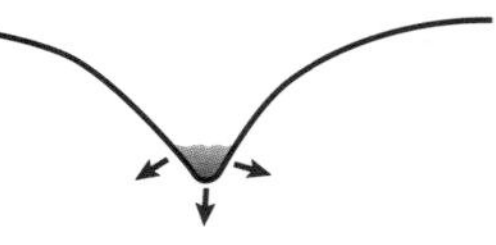

Water erodes the channel mainly downwards

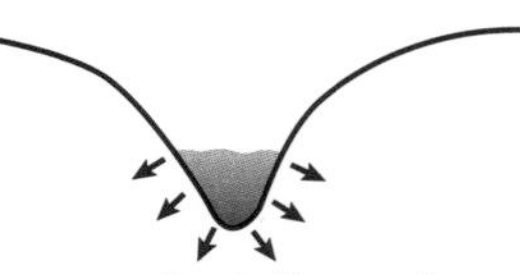

Ice erodes "all round"

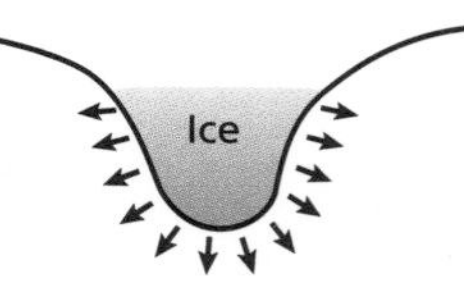

The valley becomes the channel for ice

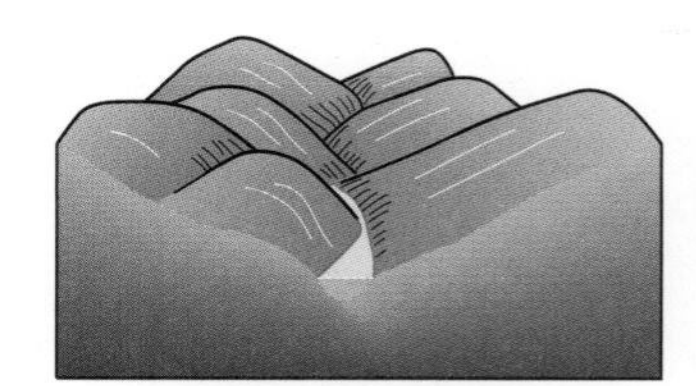

River flows between interlocking spurs

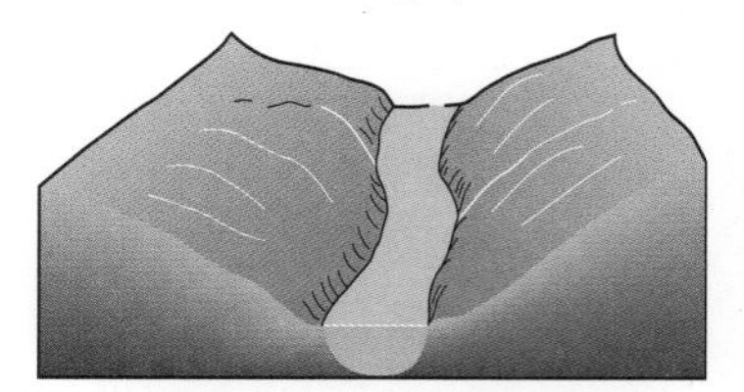

The ice carves away the land where the spurs interlocked

Landforms of Glacial Deposition

- **Terminal moraines** are crescent-shaped ridges, tens of metres high in the UK, made of unsorted material and stretching across a glaciated valley marking the furthest limit of ice.
- **Ground moraine** or **till** is a mixture of rock particles in clay smeared on to the surface and is often very thick.
- **Drumlins** are rounded, elongated, asymmetrical mounds of moraine shaped by moving ice. The 'blunt' end faces the direction of ice approach. The ice tapered the tail as it flowed over. Drumlins occur in groups.

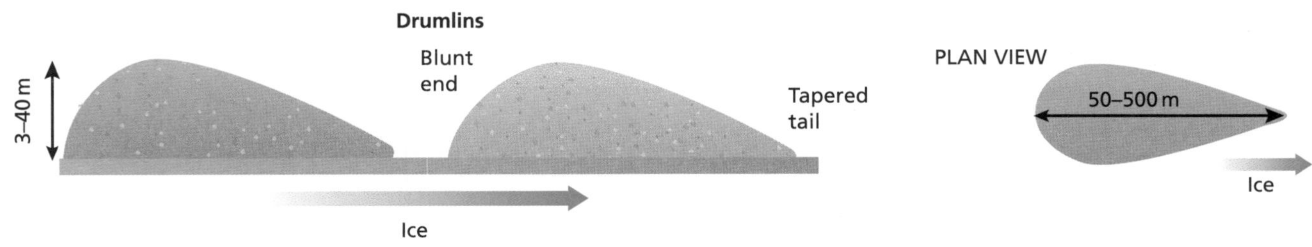

- Lateral moraines are ridges of debris along the side of the valley.
- An **erratic** is a glacially transported block left behind in an area of different rock type.
- Moraines are not solid so are easily changed by meltwater, weathering and erosion by rivers.

Key Point

Running water erosion, deposition and weathering processes change a glaciated landscape after ice has retreated.

Glaciated Upland Landscape

- Langdale Valley in the Lake District shows a range of ice-formed features and post-glacial change.

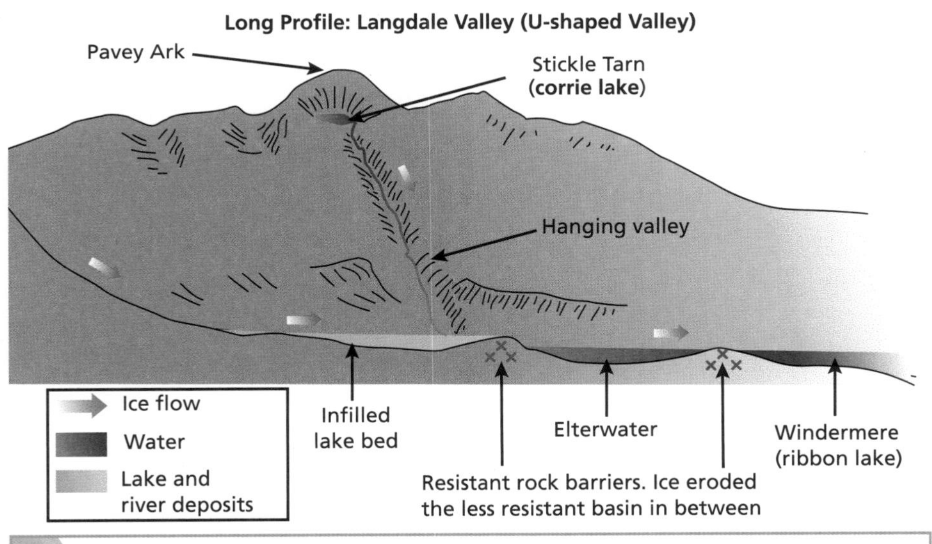

Key Words

glacial trough
U-shaped valley
trough end
truncated spurs
ribbon lake
hanging valley
corrie
arête
pyramidal peak
ground moraine
drumlin
erratic

Quick Test

1. Name two features evident along the sides of a trough.
2. Where would an arête form?
3. What would a glaciated valley floor look like immediately after the ice has disappeared?
4. What is till?

Glaciation 3: Land Use and Issues

You must be able to:
- Understand how glaciated highlands provide opportunities for, and limits to, land uses
- Describe the range of economic activities found in glaciated highland areas: industry, farming and tourism
- Explain a range of potential conflicts and management strategies.

Landscape Characteristics

- Glaciated uplands are high with steep slopes and are often **impermeable**.
- Soils are not deep, and can be **waterlogged** on flatter areas or easily **eroded** from steep slopes.
- The uplands contain different lake types (such as ribbon lakes and corrie lakes). There are many rivers.
- The uplands receive high rainfall (in excess of 1500 mm per year).
- The high altitude means that there is a short **growing season**.
- The high **terrain** means exposure to winds.
- There are low sunshine totals of under 1000 hours per year.
- High rainfall in glaciated uplands, as well as steep slopes and impermeable rock, results in rapid stream flow.
- Old **industries** used the local water supply as a raw material and power source. Nowadays, hydroelectric power (HEP) can use the fast-flowing water to generate energy.
- Lakes offer a water supply but are also a potential **flood** danger.
- Resistant rocks, such as slate and granite, provide useful resources for buildings, roofs and roads.
- Metal ores are often found in glaciated highland areas and were extensively mined.
- Nowadays the extraction of minerals attracts opposition for **environmental** and **aesthetic** reasons.
- **Pastoral farming** is more important than **arable**.
- Hardy crops, for **fodder** or **silage**, are grown on lower, flatter ground.
- **Dairy** cattle are kept on lower ground and sheep on higher ground.
- Farmland is divided into areas according to the **relief**.

Key Point

The relief and **climate** of glaciated uplands are challenging for many human activities, but can be beneficial.

Key Point

Glaciated upland areas were associated with primary industries in the past.

Land Use on a Lake District Farm

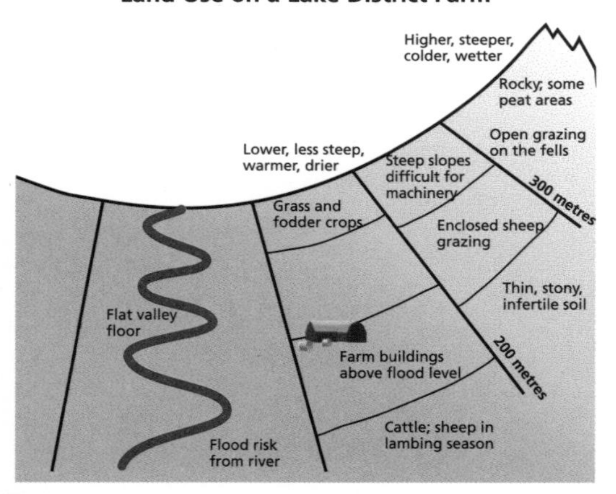

Land Use on a Scottish Upland Farm

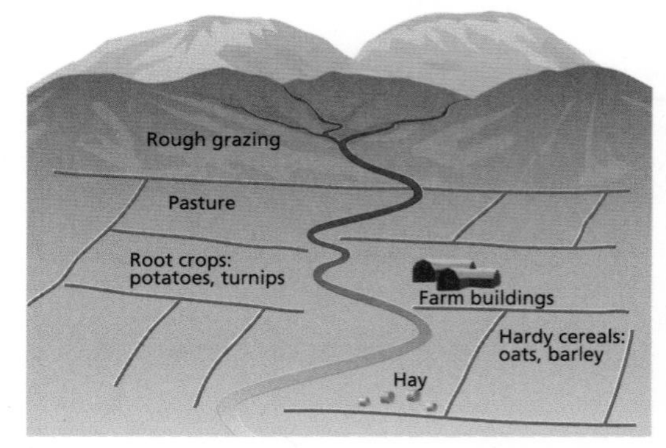

Tourism and Conflicts

- The scenery attracts visitors for a range of tourism, including both sporting and non-sporting activities.
- Farmland is often used for tourist activities.
- Diversification on farms is often related to tourist activities, e.g. the conversion of barns into accommodation.
- Old mine workings provide historical and tourist interest.
- Tourists are needed to maintain and improve the economy but can cause conflict.
- Areas such as the Lake District and Snowdonia are very popular.

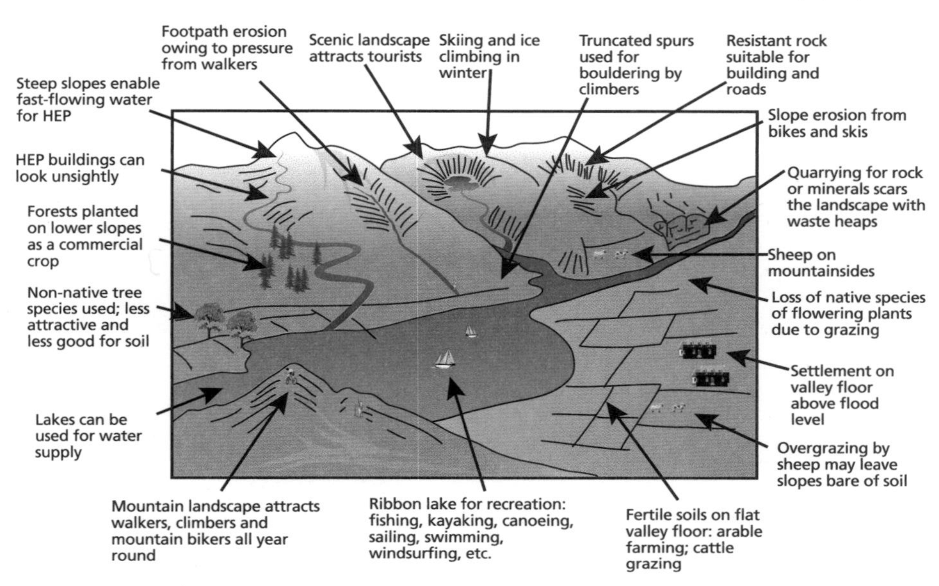

Conflicts	Examples
People and the landscape	Footpath erosion; gullying; grass or vegetation loss; litter
Tourists and locals	Crowded streets; rise in property prices due to demand for holiday lets; changing services to cater for tourists
Tourists and farmers/ landowners	Damage to crops; gates left open; rubbish in watercourses
Traffic and the landscape	Increased impermeable surfaces affects hydrology; air pollution from exhausts; road-building can change slope angles; visual pollution from proliferation of signs.

Management Strategies

- Major policies, such as the Lake District Local Development Framework 2010, can help to manage the conflicting interests and plan for a sustainable future.
- Individual strategies to help mitigate conflicts can include:
  - footpath schemes (e.g. Fix the Fells in the Lake District)
  - house-building schemes aimed at locals
  - improved bus services and cycle paths
  - provision of more litter bins
  - information boards and exhibitions provided by national parks authorities.

Quick Test

1. What is diversification?
2. Name one sporting activity associated with glaciated upland areas.
3. Give one non-recreational use (past or present) of the rivers or lakes.
4. What is the main type of pastoral farming found in glaciated uplands?

Key Words

impermeable
waterlogged
erosion
growing season
terrain
climate
industry
flood
environmental
aesthetic
pastoral farming
arable
fodder
silage
dairy
relief
tourism
diversification
pollution

Coasts 1: Processes

You must be able to:

- Describe the type and effects of waves
- Explain weathering, mass movement and erosion operating at the coast
- Describe how material is transported at the coast
- Explain why deposition takes place.

Coastal Processes and Landforms

- Factors affecting coastal processes and landforms include:
  - Waves: fetch; wave type and strength
  - Landscape processes: available material; type of erosion; amount of transport; deposition
  - The weather: rainfall; temperature change; winds; storm frequency
  - Geology: rock type; hardness; jointing; solubility
  - Biology: vegetation cover and plant root activity can damage or stabilise coastal areas; animal burrowing.

Waves

- Waves are caused by wind blowing over the ocean surface. Faster winds result in higher waves.
- Stronger waves have travelled furthest over open water (the fetch).
- Waves break in shallower water (at the plunge line) or against cliffs. The wave energy is pushed against the beach or cliff and transferred very quickly.
- Movement up the beach is swash. Return is backwash.
- Plunging: dig into the surface and remove sediment offshore.
- Spilling: long swash carrying material up a beach.
- Surging: base goes ahead and meets backwash of previous wave.
- Constructive waves: add material.
- Destructive waves: remove material; are more frequent in winter.

Plunging Wave

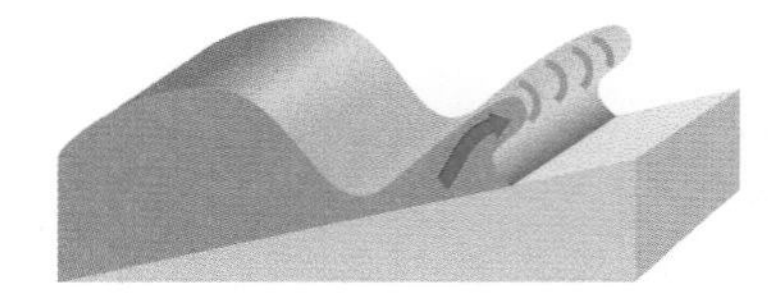

Spilling Wave

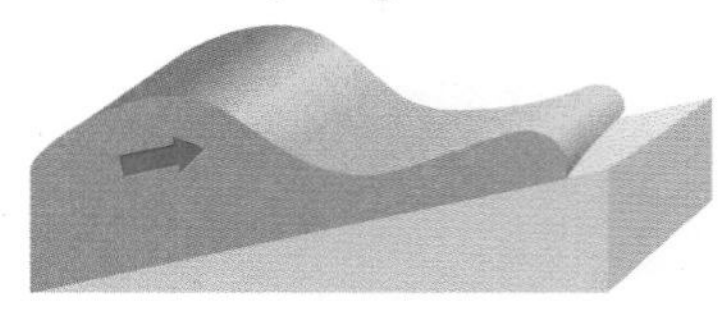

Surging Wave

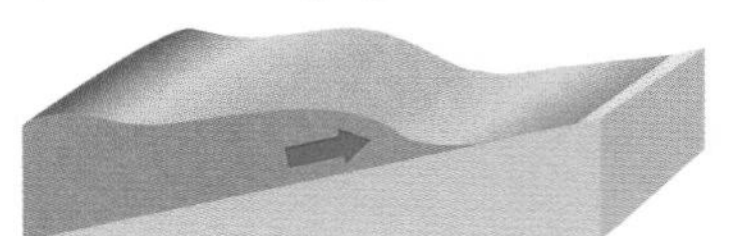

Constructive Wave

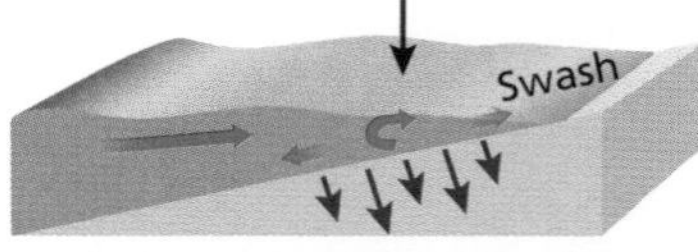

Large amount of water loss into the beach

Destructive Wave

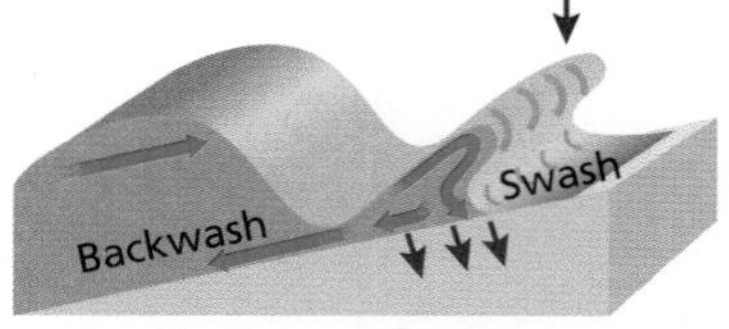

Small amount of water loss into the beach

Rock Type and Structure

- More resistant (harder) rocks, such as granite and basalt, are less easily weathered than softer rocks, such as clay.
- Higher and steeper cliffs form on harder rocks (such as limestone and granite) but cliffs are lower and rounded on clay.
- Bands of hard and soft rock parallel to the sea make a **concordant** coastline. The harder rock protects the coast.
- Rock bands at right angles make a **discordant** coastline, with headlands and bays.

Weathering and Mass Movement

- Weathering is the action of the atmosphere on rock.
- Rock is reduced in size (mechanical) and may be changed in composition (chemical).

Mechanical Weathering – Examples
• Frost action (water freezes in cracks, expands and causes rocks to break up). • Salt crystal growth (salt from seawater dries in cracks and expands each time there is wetting and drying, causing rocks to break up). • Plant roots, birds and animals burrow between joints, bedding planes and in cracks.
Chemical Weathering – Examples
• Rock decomposes and forms new materials. • Solution creates holes and pitted surfaces; important on limestones. • Acids from algae and seaweed attack rocks.

- **Mass movement** is the downslope movement of material under the influence of gravity. The material is more easily eroded by waves.

Fall

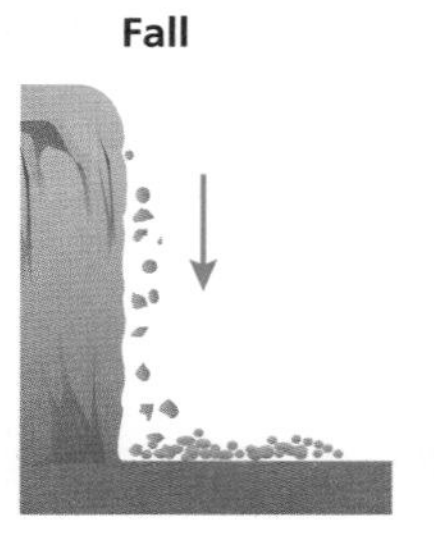

Slide

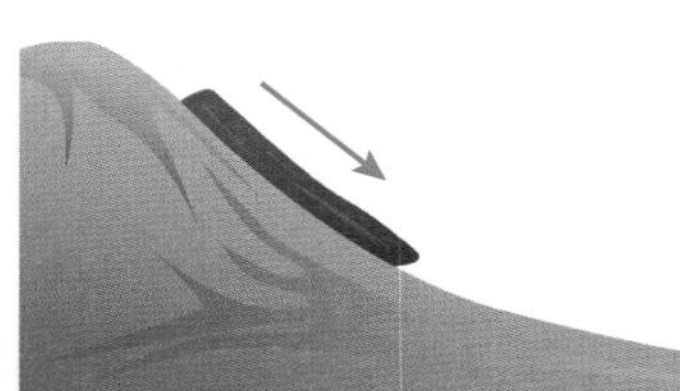

Slump

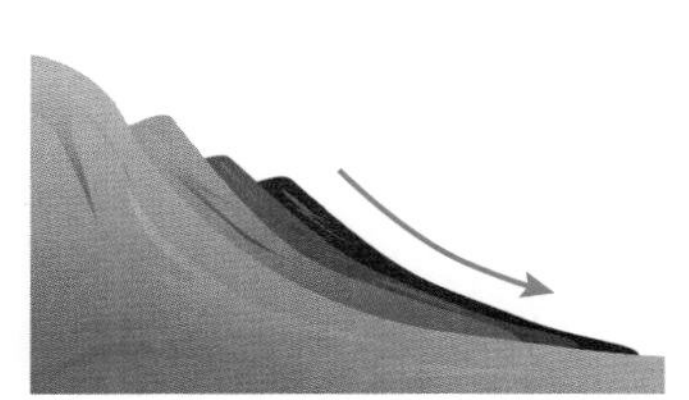

Erosion by Waves

Hydraulic Action

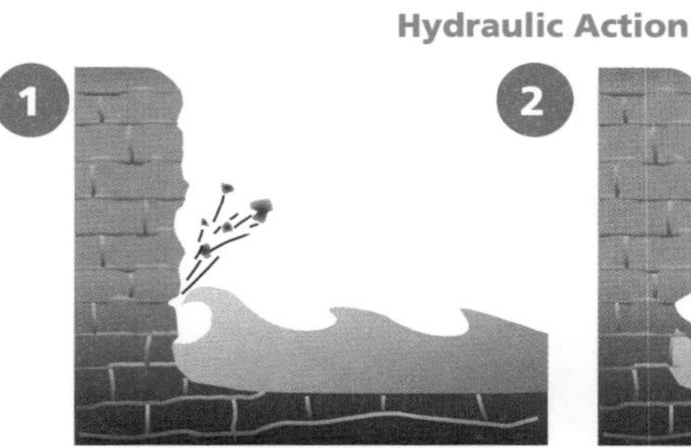

Breaking waves force air into cracks in a cliff, exerting great pressure.

Abrasion

Sand, pebbles and boulders carried by waves are hurled against cliffs, weakening and breaking the rock. Very clear on cliffs with alternating harder and softer rocks.

Attrition

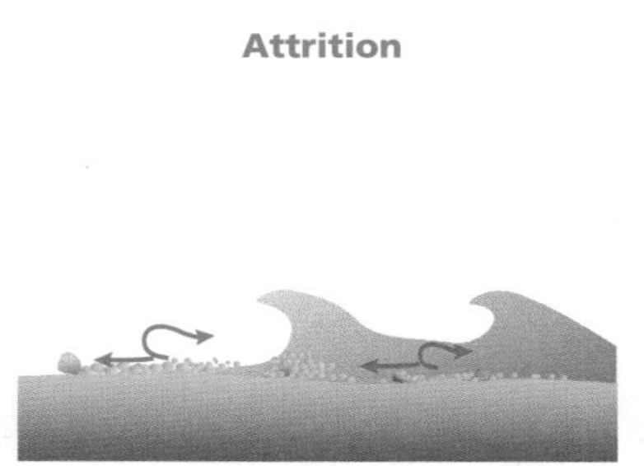

Rock pieces detached from cliffs get smaller and rounder as they crash into each other moving up and down the beach.

Transport and Deposition

- Waves move material up and down a beach by **traction** (dragging along the floor) and **suspension** (carried in water).
- Backwash brings finer material back towards the sea, contributing to surface sorting by size.
- Wind makes sand grains bounce or jump (**saltation**).
- **Longshore drift**: material moves along the beach.
- Waves deposit material when they slow, e.g. in shallower water, after breaking.
- If more sediment collects than is removed, landforms are made.

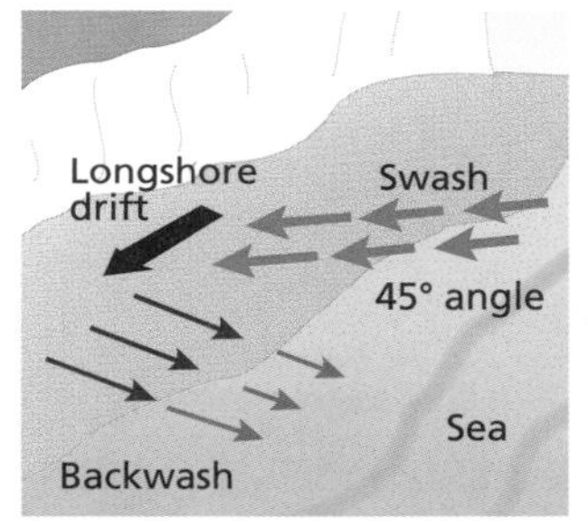

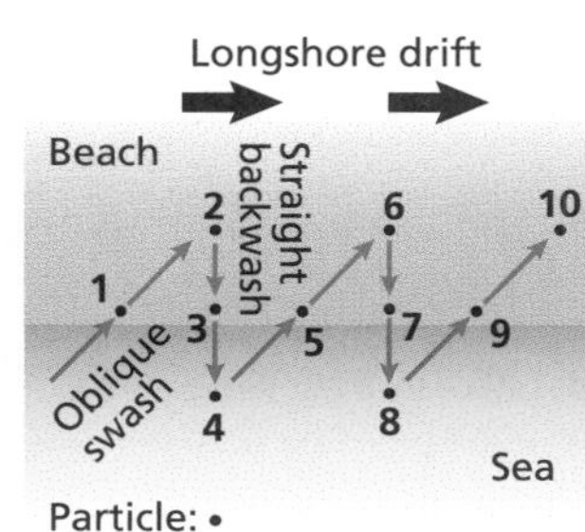

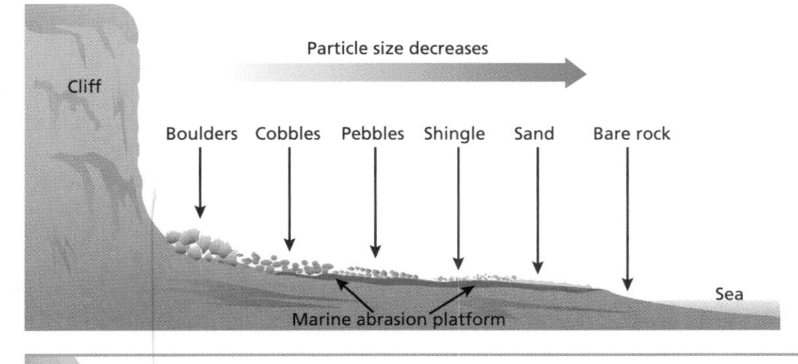

Quick Test

1. Where will beach material generally be smallest?
2. Describe cliffs that form on clay.
3. What is the fetch?
4. Which waves do most damage: spilling or plunging?

Key Words

fetch
swash
backwash
plunging
spilling
surging
constructive wave
destructive wave
resistant
decompose
solution
mass movement
hydraulic action
abrasion
attrition
traction
saltation
longshore drift

Coasts 2: Landforms

You must be able to:

- Describe headlands and bays and explain their relationship
- Explain different cliff shapes
- Understand the effects of rock type, structure, weathering and erosion on cliffs
- Describe the formation of erosional landforms: wave cut platforms, caves, arches, stacks and stumps
- Describe the formation of depositional landforms: beaches, sand dunes, spits and bars.

Headlands, Bays and Cliffs

- Wave action exploits weaknesses, e.g. faults or less resistant bands of rock on discordant coasts.
- There is deposition in bays.
- More resistant rock sticks out as headlands.
- Headlands shelter bays so material is deposited.
- Waves attack and erode headlands.
- Over time, coastlines get straighter.
- Cliffs are formed by weathering, mass movement and erosion.
- Rock type, rates of weathering and erosion are factors in the formation of different cliff shapes.

Slow Weathering

Sea more effective than weathering. Erosion and transport remove more material than inputs over time

Moderate Weathering

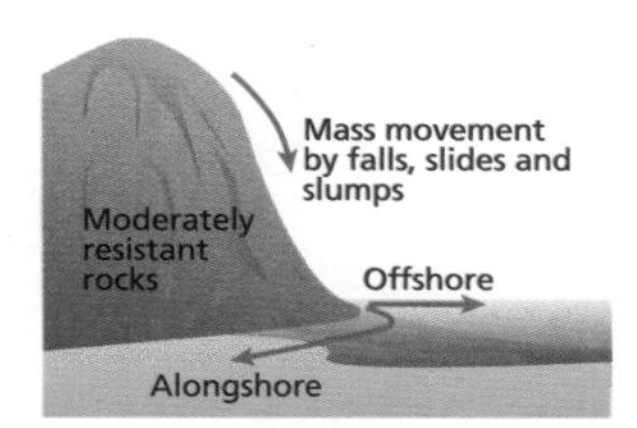

Balanced: Weathering and erosion plus transport even out over time

Rapid Weathering

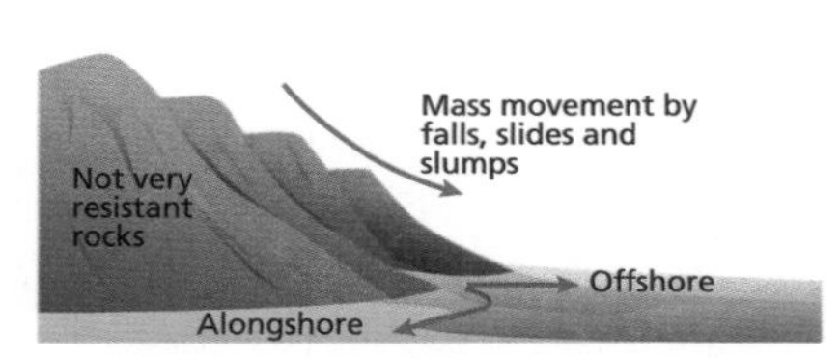

Weathering more effective than erosion and transport

Wave Cut Platforms, Caves, Arches, Stacks

- **Wave cut platforms** are gently sloping rock surfaces between the cliff base and the sea.
- Strong wave attacks create a **notch** at the cliff foot.
- Continued erosion causes cliff collapse, leaving a rock platform.
- There is usually a series of shallow steps that collect water. They are often covered with **debris**. They absorb wave energy so cliff erosion may cease.

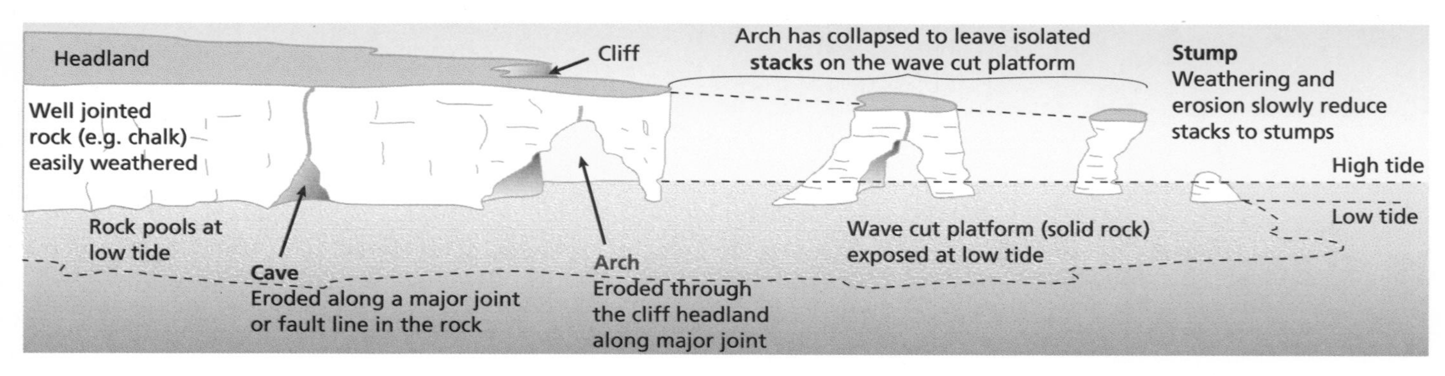

Depositional Features

- **Beaches** form from loose material, e.g. sand, shingle, pebbles.
- Sandstones create boulders; chalk forms rounded pebbles; many rocks disintegrate to sand or mud-sized particles.
- Coarser material creates steeper beaches. Fine, sandy beaches can have a 3° slope; pebbles can make a 17° slope.
- Many beaches contain pebbles deposited by glaciers.
- If the coastline is parallel to approaching waves, material is deposited as wave energy is lost by friction near the land.
- Material is brought from **offshore** and alongshore and is removed by waves and longshore drift.
- The highest level of accumulation is on confined, bay head beaches where waves quickly lose energy.
- The beach height depends on the height of waves; the beach width on the amount of sediment arriving and being removed; the beach **gradient** on the wave steepness and particle size.
- Beach profiles change in winter and summer.
- **Sand dunes** are lines of sand hills.

Key Point

The rock type is critical in the characteristics of beaches.

Small bay head beach, Cornwall

Waves rarely reach the back of a beach so the sand is dry and loose. Even between tides, it can dry enough to be bounced inland by an onshore wind.

Embryo dunes
Sand grains move up the beach by saltation and get trapped by seaweed or debris. Plants, e.g. sea twitch or lyme grass, grow and bind the sand.

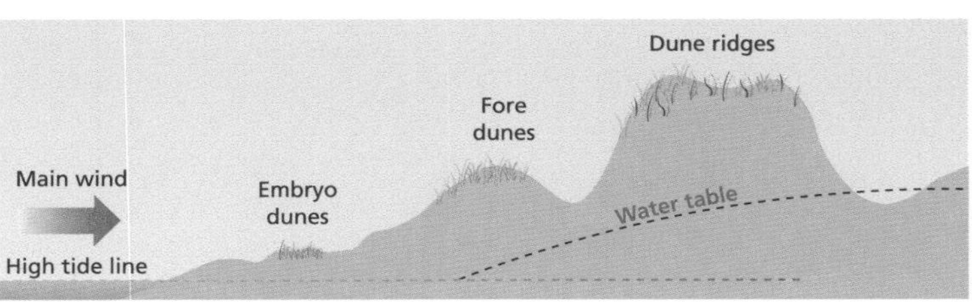

Fore dunes
Tiny dunes join up to make yellow dunes. Made stable by marram grass.

Dune ridges
Highest and biggest dunes. Look grey as plants die and decompose into the sand. Stop growing because as much sand is blown away as it arrives.

- Dunes form where there is a large store of dry sand, limited longshore drift to remove it, and consistent onshore winds.
- They are easily damaged by high winds and humans.
- **Spits** are long projections of deposited material where coasts change direction; often close to an **estuary**.
- **Bars** are sand and/or shingle ridges parallel to the coast.
- They are submerged at high tide.
- Material is deposited offshore and builds up.
- Some bars are spits that grew across estuaries. Bars can stretch between headlands with a freshwater lagoon inside them.

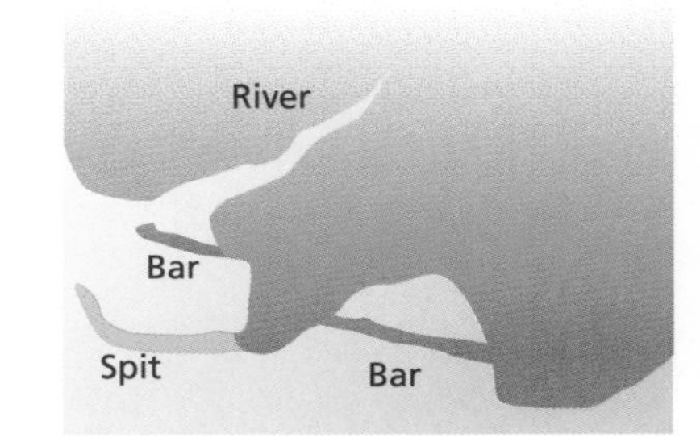

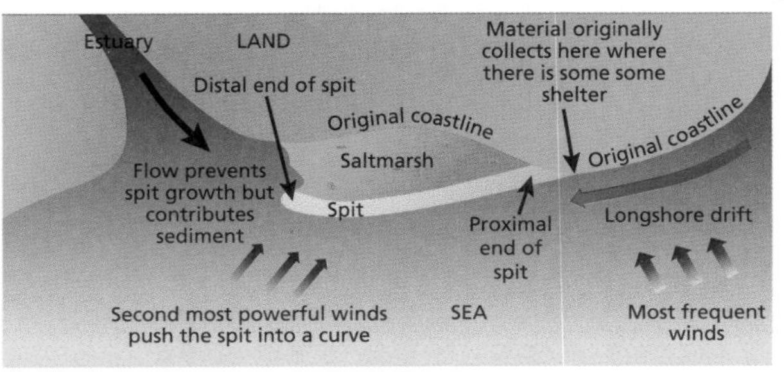

1. Longshore drift brings sediment, which collects.
2. Finer sediment is collected in calmer water behind the spit.
3. Storm waves hurl pebbles on to the spit, helping growth and stabilisation.
4. River flow stops growth.
5. Waves make the end of the spit curve inwards.

Quick Test

1. What shape are cliffs that form on very resistant rock?
2. What is the seaward end of a spit called?
3. If a stack collapses, what feature results?
4. What is a bay?

Key Words

wave cut platform
notch
debris
arch
beach
offshore
gradient
sand dunes
water table
spit
estuary
bar

Coasts 3: Management

You must be able to:

- Understand the threats to the coast
- Describe hard and soft engineering strategies
- Describe the management of an area of coastline.

Hard Engineering Strategies

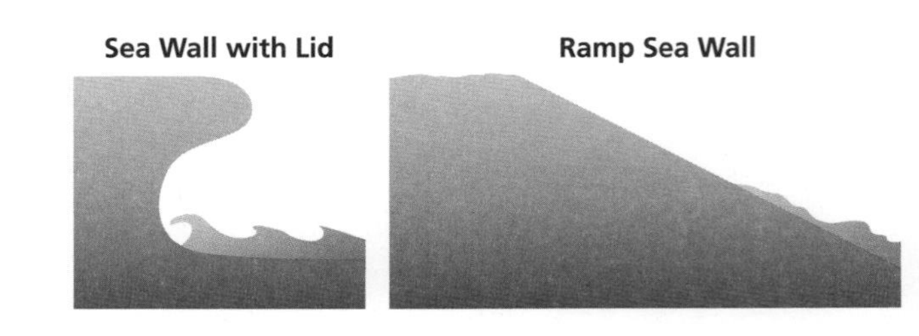

- Hard engineering strategies stop wave and tide energy transferring to land.
- They are expensive to build and maintain.
- They are very effective at protecting the land behind but not further along the coast.
- They have a great impact on natural systems.
- Cliff strategies reduce wave energy and mass movement.
- Beach management strategies aim to increase beach size.
- **Sea walls** (a cliff strategy) are built parallel to the coast and reflect – not absorb – wave energy.
- There are different styles and shapes and they are expensive to build and maintain (thousands of pounds per metre).
- Most show serious damage after 30 years.
- **Rock armour** (a cliff strategy) is also known as rip-rap.
- Rocks are pushed into and on to the cliff face.
- The rocks' size and mass absorb wave energy. Gaps trap and slow down water, so materials are less likely to be removed.
- Rocks protect against damage from particles thrown up by waves.
- Gabions (a cliff strategy) are rocks held in wire mesh structures and are usually used in conjunction with other things, e.g. walls.
- They are quite ugly and relatively cheap, with a short life-span.
- They can restrict access to beach.
- Groynes (a beach management strategy) are built perpendicular to the coast and are made of concrete, steel or wood.
- They interfere with the movement of material carried by longshore drift.
- They create wider beaches.
- If they are too short, material is taken offshore.
- Material must be spread away from groynes back on to the beach to maintain their effectiveness.

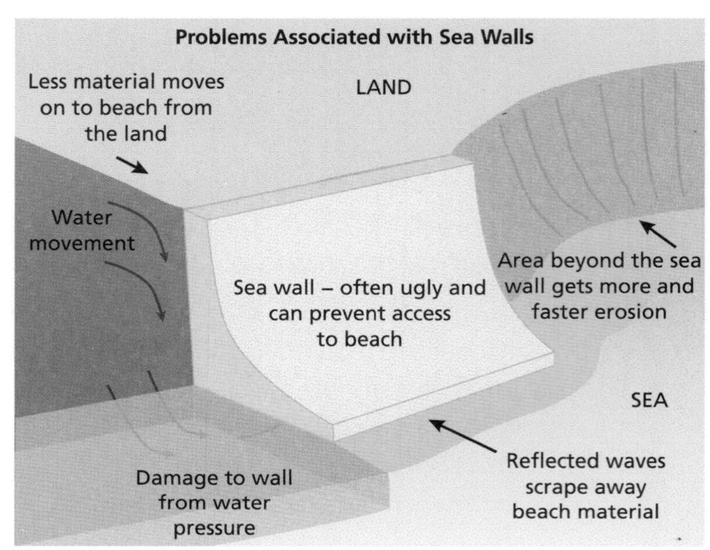

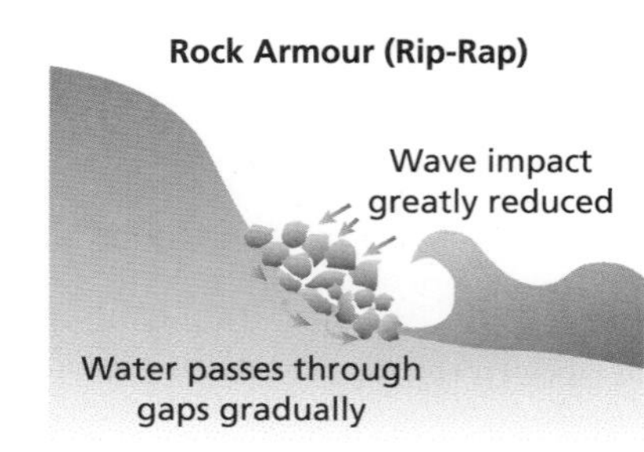

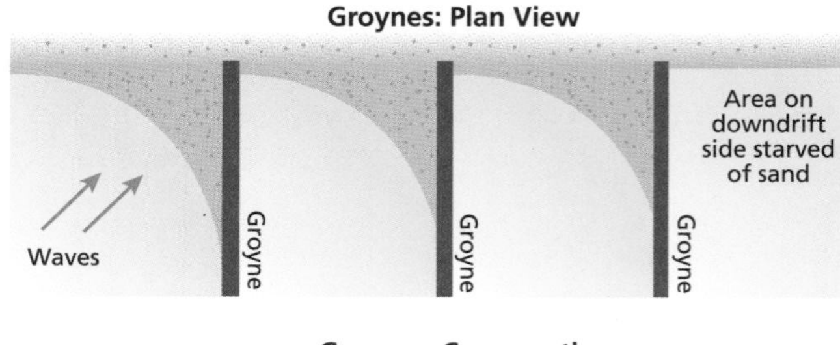

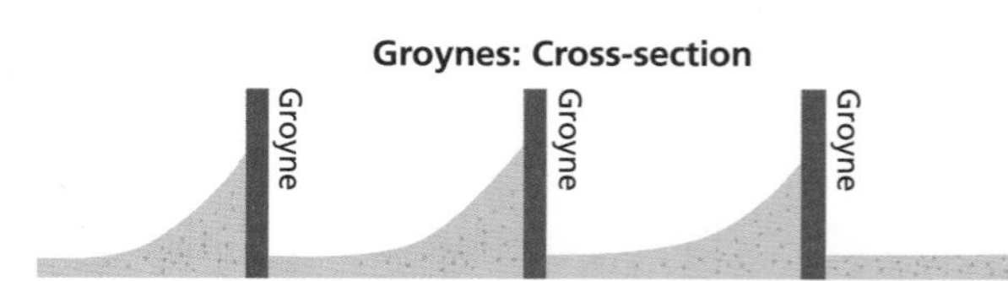

- They do not restrict access to beach.
- The places beyond groynes (downdrift) are starved of sand.

Soft Engineering

- Soft engineering works with natural processes; however, it cannot stop erosion.
- **Beach nourishment** involves adding material to an existing beach; the material should match the existing beach in size and composition.
- The material can be taken from the offshore seabed but that disturbs wave patterns.
- Wave energy is reduced.
- Moderate cost: between £5000 and £200 000 per 100 m.
- The material can be imported, but this is more expensive.
- Beach nourishment increases the attractiveness of the beach.
- **Reprofiling** involves transferring material up or down the same beach, so material automatically 'matches' the existing beach.
- The new surface is about 1 m above the highest wave run-up.
- Reprofiling can repair dune damage (blow-outs) or rebuild longer lengths of beach and it speeds up dune recovery after storms.
- Material can be taken from behind groynes or breakwaters.
- Moderate initial costs but ongoing maintenance is required, which costs thousands of pounds per 100 m.
- **Dune regeneration** is an artificial version of a natural process.
- It is low cost (£200–£2000 per 100 m) but labour intensive.
- The dunes are easily damaged in storms.
- Fences are built at the seaward edge, and sand is transferred if necessary.
- Sand accumulates around the fences.
- **Marram grass** is planted above the highest wave reach to stabilise sand. **Lyme grass** and **sea couch** are planted below.

Managed Retreat

- Managed retreat is the removal of coastal protection, usually from areas that were claimed from the sea.
- It restores original sediment movements.
- Beaches recover and grow. **Saltmarsh** regrows.
- Costs are low: only the initial removal of old structures is needed.
- It eventually creates natural defences for the **landward** area.
- An example of managed retreat is Medmerry, West Sussex (2013).

Quick Test

1. To combat which process are groynes mainly built?
2. Why are there spaces in rock armour?
3. Which are more expensive: hard or soft engineering solutions?
4. What happens to most wave energy at a sea wall?

Key Point

All forms of coastal management have benefits and costs, whether they are economic, human or landscape.

Coastal Management at Mappleton	
Location	Holderness coast, East Yorkshire (3 km south of Hornsea)
Physical background	Boulder clay cliffs easily weathered and eroded. Fastest-eroding coast in Europe. Strong north-easterly winds from the North Sea. N–S longshore drift.
Human background	Small village and farmland on the B1242 (the only direct route along the coast), 50 m from the edge of the land.
Strategies	In 1991, Norwegian granite rock armour was built along the base of the cliff and two rock groynes installed.
Results	Erosion stopped between the groynes but increased to the south.
Other points	Cost £2 million; the improved tourist facilities meant this was mainly paid for with European Union funding. In 2015, cracks on the cliffs led to a land slip, making part of the beach unsafe.

Key Words

gabions
groynes
reprofiling
marram grass
lyme grass
sea couch
saltmarsh
landward

Rivers 1: Processes

You must be able to:

- Explain the difference between a channel and a valley and name key parts of each
- Define river (fluvial) erosion, transportation and deposition
- Describe processes of erosion: hydraulic action, abrasion, solution, attrition
- Describe the ways in which material is carried by rivers
- Explain why deposition occurs
- Describe long and cross profiles of channels.

Channels, Valleys and River Systems

- A channel is the groove in which a river flows in a valley base.
- The most efficient channel is semi-circular and is twice as wide as it is deep.
- The wetted perimeter is the area touched by water.
- Water has to overcome friction to move.
- Changes to the channel affect the shape of the valley.
- The channel's **long profile** is from its source to its mouth.

State of Water in a Channel

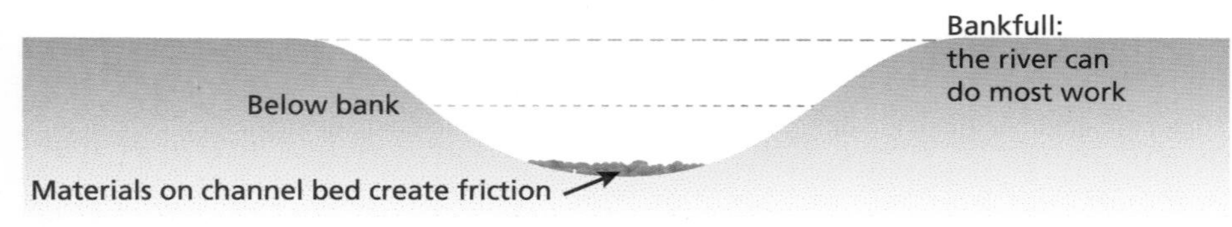

Parts of a River System

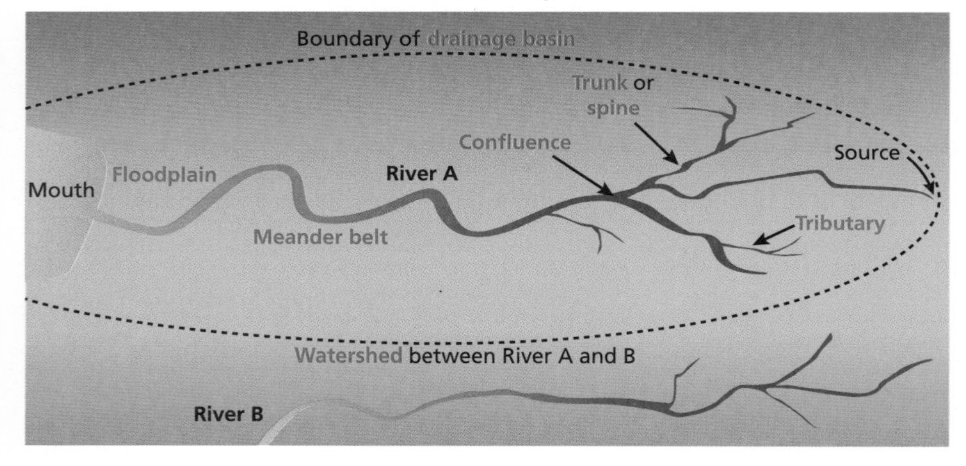

A Channel's Shape is its Cross Profile

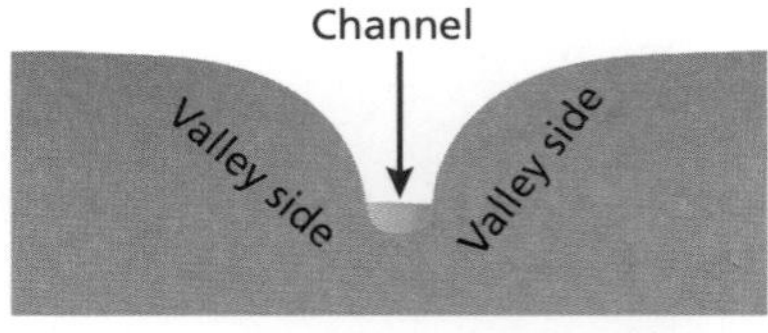

River Erosion and Processes

- River (fluvial) erosion is the wearing away and removal of material from the channel. It needs more energy than for just moving water along.
- Material needs to be entrained and moved.
- Short periods of fast flow will cause more erosion than months of lower flows.
- Vertical erosion deepens the channel; lateral erosion widens the channel; headward erosion lengthens the channel.

Hydraulic Action Attacking the Side of a Channel

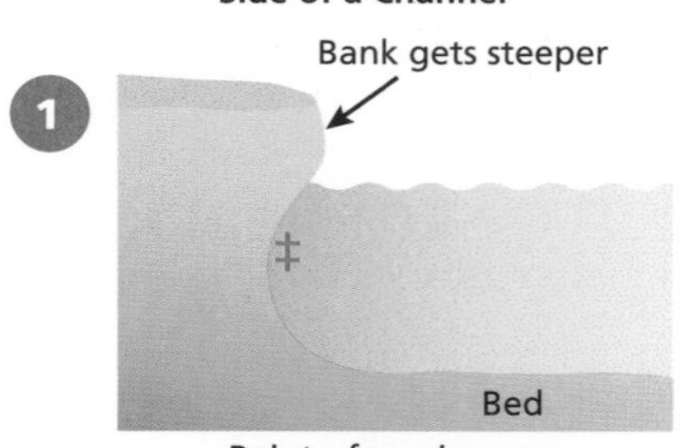

‡ Point of maximum speed and erosion

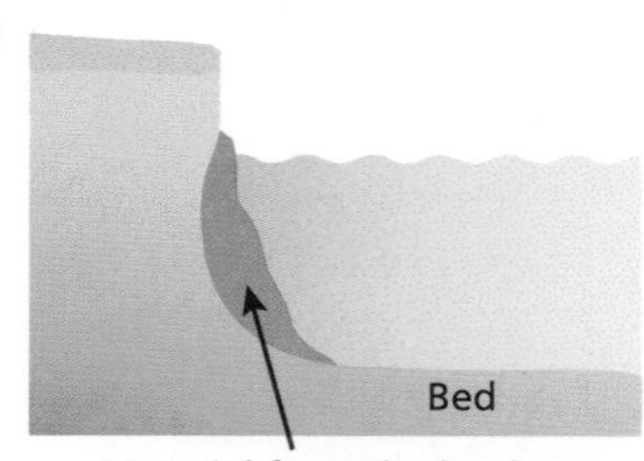

Hydraulic Action

- **Hydraulic action** is the action of moving water.
- Fast-flowing water will disturb loose material such as sand.
- River banks can be weakened and undermined, causing collapse.
- Hydraulic action is not effective on solid, hard rock.

Abrasion

- In **abrasion**, entrained material scrapes away the bed and banks.
- It can chip away at solid rock, creating small particles that can be carried along, and can smooth the bed and banks.
- Most downcutting of a channel is the result of abrasion.

Solution

- Dissolved material is carried away in **solution**. This is most effective on limestones but almost all rocks are partially soluble.

Attrition

- **Attrition** results from load particles crashing into each other.
- Particles being carried along get smaller and rounder.

Key Point

Coarser material like sand is easier to entrain but more difficult to carry; finer material like clay is harder to entrain but easier to carry.

Key Point

Attrition affects the load, not the channel, but it makes material easier to carry and to be used for abrasion.

Transportation and Deposition

- Material that has been entrained and then carried (**transported**) by a river is called its load.
- Faster flow enables more and bigger particles to be carried.
- Material carried in the body of the water is the suspended load.
- Material carried along the channel bed is the bedload.
- Bedload is moved by **traction** and **saltation**.
- Soluble material is carried in solution and is invisible.

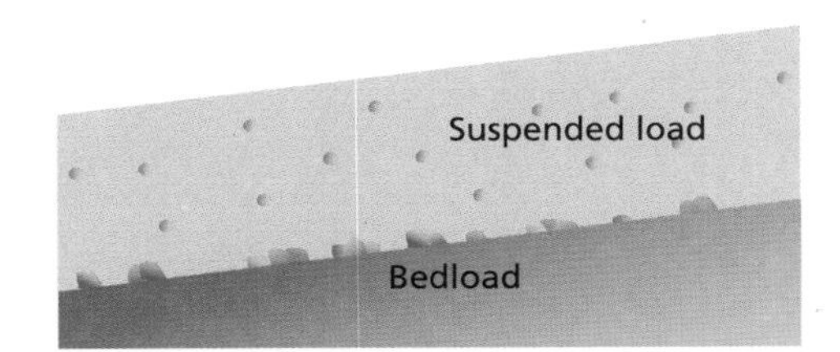

- **Deposition** can occur at any point along the length of the river.
- It can be found in the same places as where erosion is taking place.
- Deposition occurs because there is not enough energy in the water to transport material.

Quick Test

1. What is material carried by a river called?
2. What effect can attrition have on rock particles?
3. What does lateral erosion do to the channel?
4. Where can deposition happen?

Key Words

channel
valley
wetted perimeter
source
mouth
confluence
spine
trunk
drainage basin
meander belt
tributary
floodplain
watershed
fluvial
entrained
headward
river bank
undermined
downcutting
limestones
load
suspended load
bedload

Rivers 2: Landforms

You must be able to:

- Describe interlocking spurs, waterfalls, gorges, meanders, ox-bow lakes, floodplains, levees and estuaries, and explain their formation
- Understand the varying roles of erosion and deposition in making landforms
- Understand the relationship between landforms in a river landscape.

Interlocking Spurs

- Most river energy is used to overcome friction and move water.
- When flow is strong, small and highland streams mainly erode downwards by hydraulic action.
- Water flows around obstructions as it cannot wear through them.
- Land sticking into the route of the channel forms interlocking spurs that restrict views up or down a valley.

Interlocking spurs on a Pennine stream

Waterfalls

- Waterfalls are a sudden steepening in the course of a river.
- They form where there are bands of rock with different hardnesses, faulting, at plateau edges and with deposition in a channel.
- In waterfalls, water is not slowed by friction so deepens the pool at the front of the fall and the rock behind it is eroded by hydraulic action and abrasion.

Waterfall Associated with a Fault

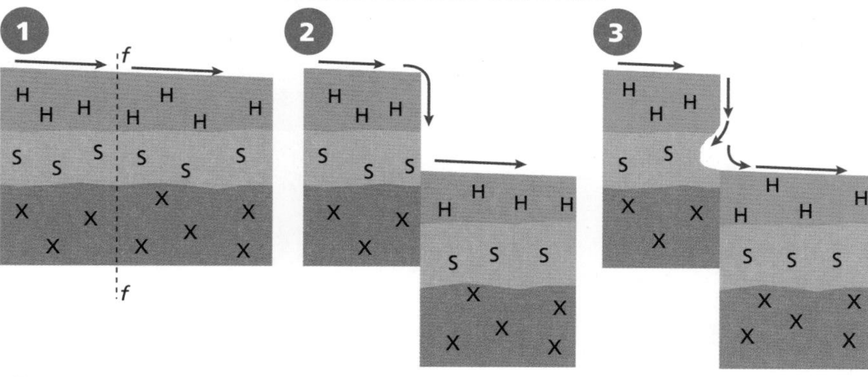

Waterfall Associated with Horizontal Bands of Rock of Different Hardness

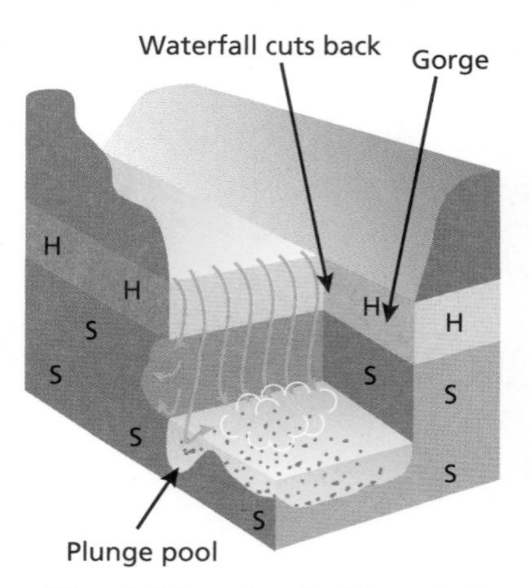

Waterfall Associated with Vertical Bands of Rock of Different Resistance

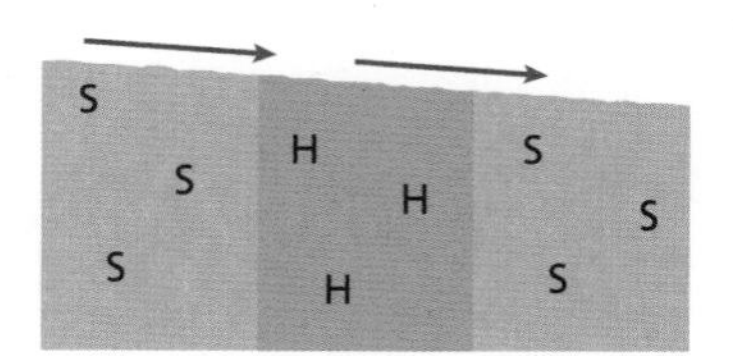

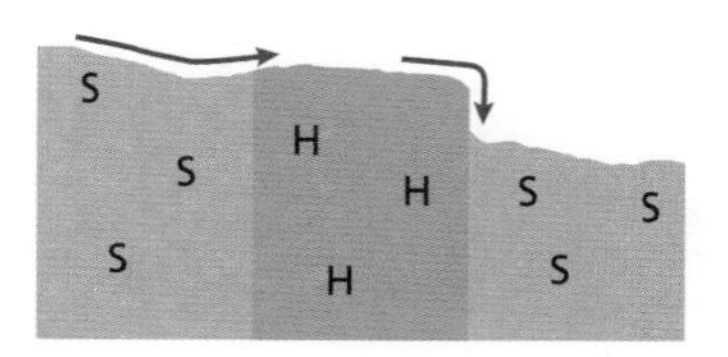

Gorges

- Gorges are steep-sided sections of a valley downstream from a waterfall.
- They are the result of waterfalls eroding backwards into the rock.

Meanders

- Meanders are well-developed river bends. The most efficient route in a channel is the thalweg, which is sinuous.
- Water travels at different speeds.
- Faster water hits the outer bend, abrading the bank and making steep river cliffs. Flow at the outer bend is helicoidal.

Key Point

Some features are a result of both erosion and deposition.

- At inner bends, water is slower and deposition takes place; a slip-off slope forms.
- Meanders have an **asymmetric cross-section**.
- Over time, meanders move sideways and downstream.

Helicoidal Flow – Cross-section

The water flows downstream from outer to the next inner bend
Water level
River cliff
Slip-off slope
Convex inner bend
Concave outer bend

Ox-bow Lakes

- Ox-bow lakes are cut-off meanders.
- Meanders become increasingly developed, forming a narrow neck. At times of very high water flow, the meander neck is weakened, eventually breaking through.
- Water flows through both the meander and the straightened part but deposition seals off the meander, leaving an ox-bow lake.

Floodplains

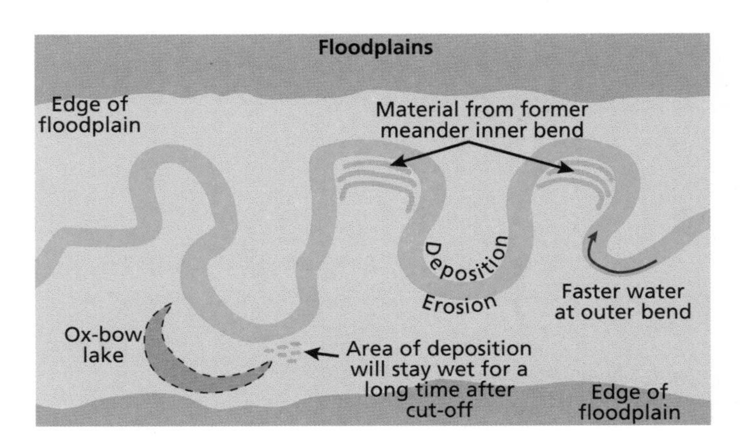

- Floodplains form at the side of a river channel. They are made from alluvium – material carried and deposited by rivers.
- Some material may be from rivers overflowing their banks.
- The water not held in a channel spreads and slows, so even very fine material is deposited. Coarser material comes from the inner bends of meanders as they move sideways and downstream.

Levees

Levees

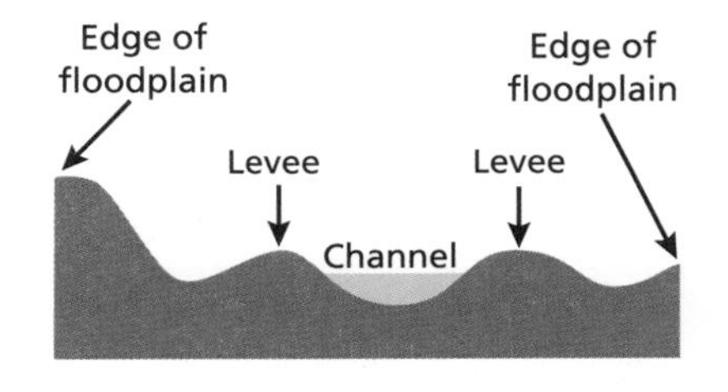

- Levees are raised areas of coarser material, beside a river channel.
- At times of flood, water leaves the channel, suddenly slows and deposits material. The river bank is raised.

Estuaries

- **Estuaries** are tidal river mouths, found on rivers with wide mouths.
- River sediment flocculates (clumps together) when freshwater meets saltwater, and sinks. Mudflats form.
- Estuary material is generally finer towards inner areas and coarser at outer edges.

Key Point

Familiarise yourself with the landforms evident on the long profile of a UK river (e.g. the River Tees).

Quick Test

1. What is the steep-sided feature found in front of a waterfall?
2. What is the inner bend of a meander called?
3. What is most energy in a river used for?
4. What is alluvium?

Key Words

friction	gorge	ox-bow lake
interlocking spurs	meander	alluvium
waterfall	thalweg	levee
faulting	sinuous	tidal
plateau	helicoidal	flocculates
	slip-off slope	mudflats

Rivers 3: Flooding and Management

You must be able to:

- Explain how water gets into a channel
- Understand why channels fill more quickly or slowly
- Extract information from hydrographs
- Understand why floods occur
- Explain how floods can be prevented or their impacts reduced.

How Water Gets into a Channel

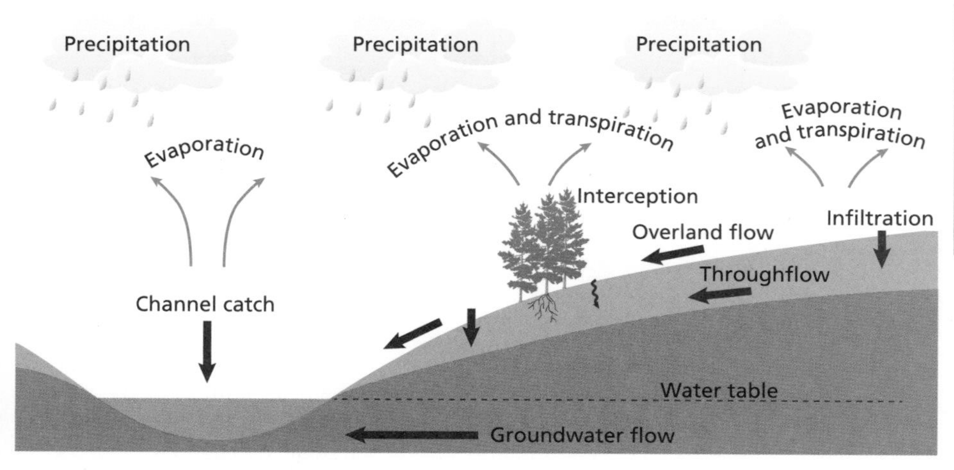

	Where from	Quick or slow?
Channel catch	Rain going directly in	Quick
Overland flow	Valley side surfaces	Quick
Throughflow	Soil under valley sides	Slow
Groundwater flow	Rocks below the river	Very slow

- Channels fill quickly if valley sides are **steep** (relief); if valley sides are impermeable (**geology**); if there is nothing to slow the water, e.g. trees (land use); if soils are already **saturated** (existing conditions).

Discharge and Hydrographs

- Discharge is the amount of water in a channel that passes a place at one time. It is measured in **cumecs** (cubic metres per second) and is calculated from the cross-sectional area of the water and its velocity.
- **Hydrographs** show river discharge over time.
- Steep rising and falling limbs mean discharge increases then decreases quickly – a **flashy** river.
- Gentle rising and falling limbs mean discharge changes slowly – a **subdued** river.
- **Lag time** is between peak rainfall and peak discharge.

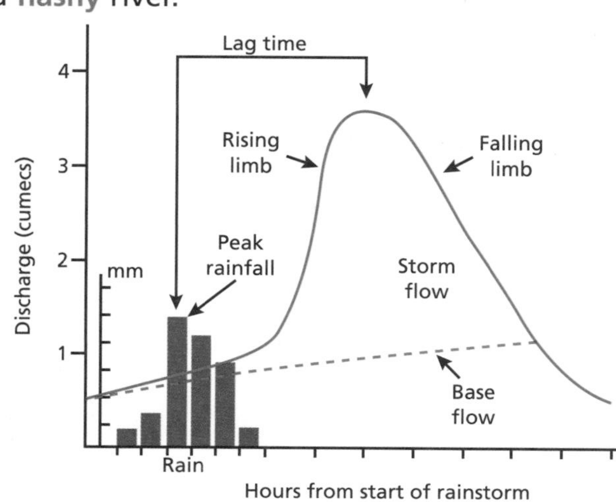

KEY
...... Wetted perimeter
d depth
w width
v velocity

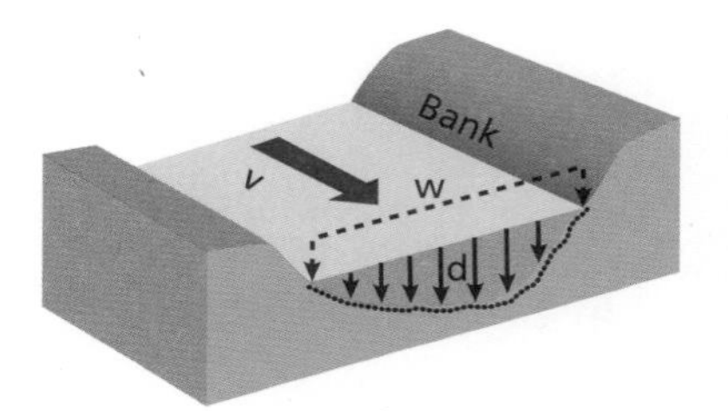

Cross-sectional area = w × d
Discharge = w × d × v

Contrasting Hydrographs

River A: flashy
River B: subdued

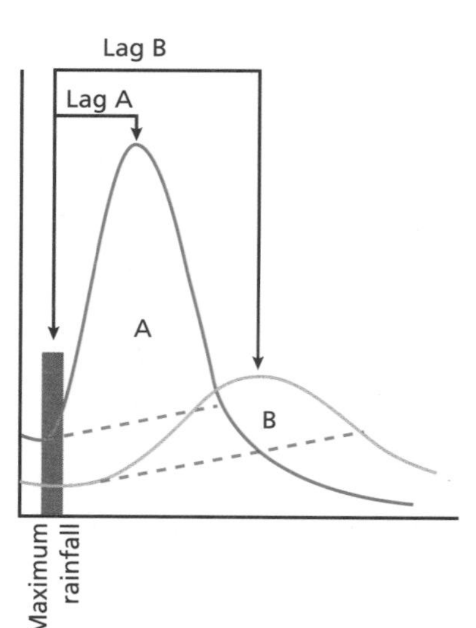

Floods

- Most British rivers overbank and flood once every 2–3 years.
- Channels are adjusted to 'usual' conditions.
- Water level rise can be dramatic, e.g. the River Valency in Cornwall rose 2 m in just under an hour on August 16, 2004.
- Causes of floods are: sudden, heavy storms; long periods of continuous rain; slow-moving storms; melting snow and ice; high tides; human factors.

Preventing Floods

- The aim is to slow down water getting into the channel (abatement) and/or speed water through the channel.
- **Hard engineering:**
  - Artificial structures.
  - Expensive.
  - Intrusive on the landscape.
  - Slow to build but have an immediate effect.
- **Soft engineering:**
  - Working with the river landscape.
  - Cheap; often zero cost.
  - Can be slow to take effect.
  - In tune with the landscape.

Hard Engineering Strategies	Soft Engineering Strategies
Straightening channels increases water velocity.	Trees intercept rainfall and take up water.
Dredging increases channel size but reduces speed, so more deposition.	Terracing valley sides slows overland flow.
Retention ponds allow water to go overbank into specific places.	Removing boulders speeds up flow and are then used to raise and strengthen banks.
Floodways alongside rivers take flood waters from the channel.	Restricting building on floodplains reduces impermeable surfaces.
Dams regulate flow.	Flood warnings, risk maps, preparation advice from the Environment Agency.
Embankments raise channel sides but move the danger downstream.	Earth embankments use local materials and blend into the landscape but can get waterlogged or destabilised by plants and animals.
	River restoration puts river back to its natural state.

York – A Complex Situation

- Low annual rainfall: 600 mm per year.
- Low-lying: 15 m above sea level.
- The rivers Ouse and Foss flow through the city. The rivers Swale, Ure and Nidd flow from the Yorkshire Dales into the Ouse.
- Changed land use, improved pasture and more building in the Yorkshire Dales speeds water into rivers.
- The North York Moors is overgrazed, has peat eroded by walkers and has lost forests, so water flows faster into the city.
- York is full of impermeable surfaces and has seen more and more house building.

York – Flood Defences

- Earth embankments around properties at the city edge.
- 'Washlands': store water from the Ouse when it rises.
- Concrete embankments and flood walls along river sides.
- Foss barrier: lowered to keep Foss water back when Ouse is high.
- The sunken gate gets raised to link city walls with flood barrier.
- Businesses and houses clear cellars to hold flood waters.

Quick Test

1. Which measurements are needed to calculate discharge?
2. What is lag time?
3. Give an example of an impermeable surface.
4. What is throughflow?

Key Words

channel catch
overland flow
throughflow
groundwater flow
steep
geology
saturated
cumecs
hydrograph
flashy
subdued
lag time
flood
intercept
dredging
terracing
embankment
peat

Review Questions

Ecosystems and Balance

1. Give an example of a global large-scale ecosystem. [1]
2. Define the term 'producer'. [2]
3. Define the term 'consumer'. [2]
4. Define the term 'decomposer'. [2]

Total Marks .... / 7

Ecosystems and Global Atmospheric Circulation

1. Which line encircles the Earth and divides it into the Northern Hemisphere and the Southern Hemisphere? [1]
2. Which line encircles the Earth at roughly 23° **north** of the Equator? [1]
3. Which line encircles the Earth at roughly 23° **south** of the Equator? [1]
4. Using examples, describe the global distribution of the tundra ecosystem. [4]

Total Marks .... / 7

Rainforests and Hot Deserts – Characteristics and Adaptations

1. In tropical rainforests, where are the majority of nutrients found? [1]
2. What word describes the artificial watering of crops? [1]
3. How have pitcher plants evolved to cope with the low nutrient soils in rainforests? [2]
4. Describe the climate in hot deserts. [2]

Total Marks .... / 6

Opportunities, Threats and Management Strategies in the Amazon

1. What do the initials CITES stand for? [1]
2. What happens to the soil after vegetation cover is removed? [1]
3. What is the difference between commercial and subsistence farming? [2]
4. Using examples, describe how a named HEP scheme can have both benefits and drawbacks. [4]

Total Marks .... / 8

Opportunities, Threats and Management Strategies in Hot Deserts

1. Name one HEP scheme in the Mojave Desert. [1]
2. The mining of which mineral has led to a reduction in underground aquifers? [1]
3. What are 'terraced' fields? [2]
4. Describe how the problem of desertification can be tackled in southern Spain. [4]

Total Marks .... / 8

Polar and Tundra Environments

1. What word describes the characteristic landscape type in tundra regions? [1]
2. Name two international agreements that control the human use of Antarctica. [2]
3. Using examples, describe how animals have adapted to life in polar and tundra climates. [4]
4. Name three resources found in and around Antarctica. [3]

Total Marks .... / 10

Practice Questions

Glaciation 1: Processes

1. What is the material called that is pushed along by a glacier and deposited at its furthest point? [1]
2. Where does abrasion take place? [2]
3. On the diagram below, insert an arrow to indicate the direction of ice flow and label these in the correct places: Bedrock; Plucking; Debris; Plucked block; Abraded surface. [6]

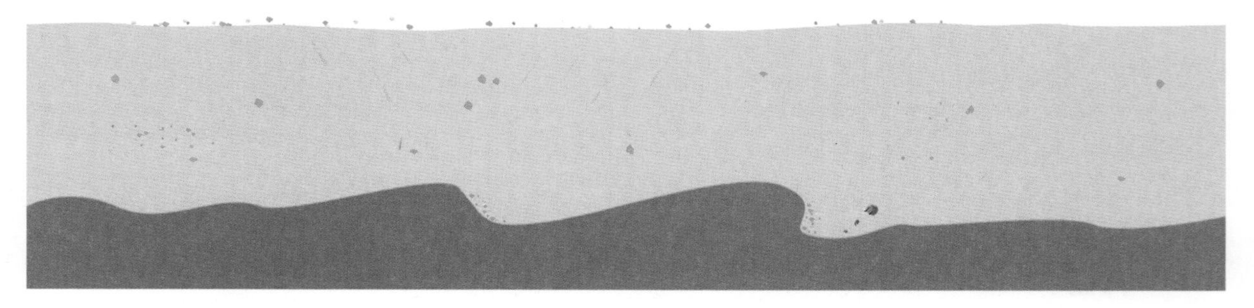

Total Marks .... / 9

Glaciation 2: Landscape

1. Where is the trough end in a glaciated valley? [1]
2. Describe the appearance of the back wall of a corrie. [2]
3. Why might a glaciated valley be flat and quite smooth today but was rocky just after the ice disappeared? [4]

Total Marks .... / 7

Glaciation 3: Land Use and Issues

1. Why is flooding a potential risk in glaciated areas? [4]
2. How do glacial landforms favour recreation activities? [4]

Total Marks .... / 8

Coasts 1: Processes

1. Explain why beach material gets smaller and rounder as you move from a cliff to the sea. [2]
2. Explain why different rock types produce different shapes of cliff. [4]

Total Marks .... / 6

Coasts 2: Landforms

1. Look at the photograph of part of the Shiant Islands in Scotland.
   a) What is feature A? [1]
   b) With continued erosion, what feature might be formed in the same place? [1]
2. Why might a spit have a curved end? [2]
3. What conditions are best for the creation of sand dunes? [3]

Total Marks .... / 7

Coasts 3: Management

1. Explain **one** negative point about sea walls as coastal protection. [3]
2. Compare beach nourishment and beach reprofiling. [4]

Total Marks .... / 7

Practice Questions

Rivers 1: Processes

1. On what type of rocks is solution most effective? [1]
2. Describe one way in which bedload can be moved. [3]
3. Explain the connection between entrainment and erosion. [3]

Total Marks .... / 7

Rivers 2: Landforms

1. Look at the photograph of a river in Scotland.

   a) Name the features labelled A and B. [2]

   b) What feature is likely to form at C? [1]

   c) What part of the river is at location X? [1]

   d) What process is operating at location X? [1]

2 Draw the cross-section of a channel at a meander bend.
Add these labels: Inner bend; Outer bend; Deposition; Erosion. [4]

Total Marks .... / 9

Rivers 3: Flooding and Management

1 Give two reasons why a river might be 'flashy'. [2]

2 Look at the storm hydrograph below.

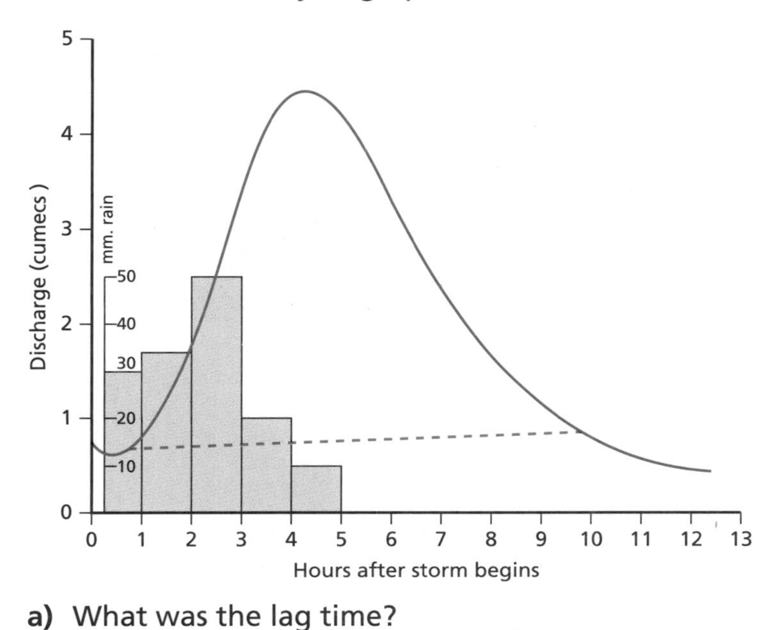

a) What was the lag time? [1]

b) What was the highest rainfall? [1]

c) What was the highest discharge? [1]

d) For how many hours was the river above normal flow? [1]

3 Describe one **slow** way in which rainwater can reach a channel. [4]

Total Marks .... / 10

Urbanisation

You must be able to:

- Understand how rapidly the world's urban population is growing
- Explain the reasons why people migrate to urban areas.

Urbanisation

- Urbanisation means an increase in the proportion of people living in urban (town and city) areas.
- In 1900, 10% of the world's population lived in urban areas. This had risen to 34% by 1960 and to 54% by 2015. The United Nations predicts that by 2030, 60% of the world's population will be urban.
- Levels of urbanisation are increasing across the world, with the fastest rates of growth occurring in lower income countries (LICs).
- As countries industrialise, levels of urbanisation increase as people move into cities for jobs.
- Most higher income countries (HICs) industrialised over 100 years ago and now have large urban populations. LICs are still in the early stages of industrialisation and so their urban populations are still growing.
- As urban areas grow outwards, they encroach into rural (countryside) areas. This is known as **urban sprawl**.

Key Point

The majority of the world's population now live in urban areas.

World Mega-Cities

- A millionaire city is one with a population of over 1 million people. There are currently 280 millionaire cities in the world, and the number is growing rapidly – mostly in LICs.
- A city with over 10 million people is known as a mega-city. There are 35 of them in the world, mainly in LICs and newly emerging economies (NEEs).
- LIC cities are growing so fast due to rural to urban migration and high natural increase in population (high birth rates and falling death rates).

Key Point

Most of the world's largest cities are in LICs.

Rural to Urban Migration

- Push factors are those that drive people away from rural areas, e.g.:
  - unemployment
  - low wages
  - drought, famine and other natural disasters
  - farming is difficult and unprofitable
  - few job opportunities
  - lack of social amenities
  - isolation.
- Pull factors are those that attract people into cities, e.g.:
  - more job opportunities
  - higher wages
  - better schools and hospitals
  - better housing and services (like water, electricity and sewerage)
  - better social life
  - better transport and communications.

City life can attract people from rural areas

Key Point

LIC cities are growing rapidly, mainly due to rural to urban migration.

Quick Test

1. What is a millionaire city?
2. What percentage of the world's population was urban in 2015?
3. How many people are needed to qualify a city as a 'mega-city'?

Key Words

urbanisation
millionaire city
mega-city
push factors
pull factors

Urban Issues and Challenges 1

You must be able to:

- Explain how urban growth creates opportunities and challenges within cities in LICs and NEEs
- Explain different ways conditions in shanty towns can be improved for the residents.

Case Study: City in a Newly Emerging Economy (NEE) – Rio de Janeiro, Brazil

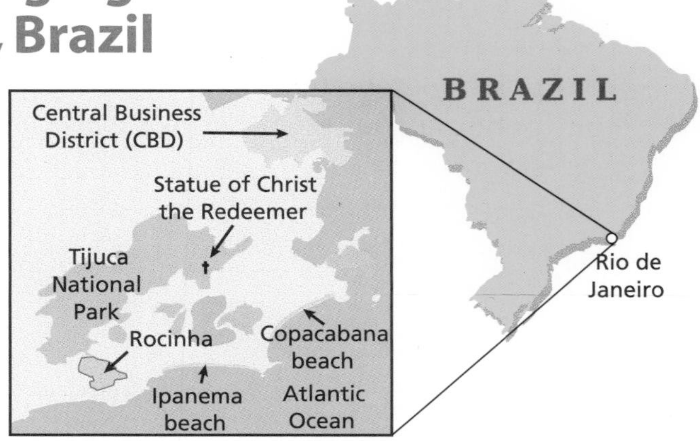

- Population: 12 million (2017).
- 23% of population live in over 600 favelas.
- Location: south-east coast.
- Reasons for growth: rural to urban migration and natural population increase.
- Former capital city of Brazil.
- Major tourist attractions: Copacabana Beach, statue of Christ the Redeemer, Sugar Loaf Mountain.
- Hosted football's World Cup in 2014 and the Olympic Games in 2016.

Almost a quarter of Rio's population live in favelas.

Challenges in Rio

Social	**Migration:** Rapid growth in recent years because migration from rural areas to the city has put huge pressure on services and amenities. Migrants have been attracted to Rio due to job opportunities, higher wages, and better services like schools and hospitals.
	Housing: Rio has a number of shanty town settlements (favelas) that house the many rural migrants who come to the city. They are unplanned and spontaneous – often growing up on poor quality land. Residents have no legal land ownership. Houses are built using cheap materials like wood or corrugated iron and many people are crammed into a small area, leading to overcrowding, with no clean running water or sewage disposal. There are no schools or hospitals, few job opportunities and high levels of disease and illness.
	Healthcare: Has been poor but more clinics have now been established, especially to improve maternity care and support for the elderly.
	Education: Schools suffer from low enrolments, and funding has been made available for higher pay and better training for teachers, as well as grants for poor families.
	Water supplies: A clean water supply is not available for 12% of the population. New treatment plants have been built and new pipelines laid.
	Energy: A shortage of electricity has meant frequent blackouts. New power lines have been constructed, along with a nuclear power station and a new HEP station.
	Crime: Street crime has been high, with powerful drugs gangs controlling the favelas. The police have taken steps to reduce crime, including the 'Pacifying Units' that have reclaimed order in some of the favelas.
Economic	**Poverty:** There is a huge gap between rich and poor citizens in Rio. Some favelas have grown on hillsides right next to the affluent central business district.
	Employment: Unemployment remains high in the favelas (over 20%), where most people work in the informal economy. Poor transport systems make it difficult for favela dwellers to access other parts of the city.
Environmental	**Urban sprawl:** As the city continues to grow, it encroaches on surrounding rural areas.
	Pollution: Air pollution from heavy traffic and congestion is a major issue, as is pollution of the sea from sewage and industrial waste.
	Waste disposal: This is a particular problem in the favelas, many of which are inaccessible to collection vehicles.

Opportunities in Rio

Social	**Cultural and ethnic diversity:** Rio is home to a huge mix of different races, religions and cultures. The annual Rio Carnival is an international event.
	Education: Rio has a number of universities and is a centre for research and development in Brazil.
	Strength of community: There are many positive aspects to life in favelas, including the people who create their own economy, e.g. shops, restaurants and cottage industries (e.g. pottery); residents send money back home to families in rural villages; recycling commonly takes place, for example of waste and building materials.
	Transport: Economic development has led to improvements in roads and transport systems.
Economic	**Industry:** Industrial growth has boosted Rio's economy through steel making, port industries, oil refining, petrochemicals, manufacturing (including modern industries like computers and electronics) and a growing range of services such as banking and finance.
	Tourism: Rio is one of the most-visited cities in the Southern Hemisphere. Major attractions include Copacabana Beach, Ipanema Beach, the statue of Christ the Redeemer and Sugar Loaf Mountain.
Environmental	**Beaches:** The Atlantic beaches continue to attract tourists.
	Forests: The Tijuca National Park is one of the largest urban forests in the world.

Rocinha Shanty Town, Rio de Janeiro

- The rapid growth of Rio's population has led to a severe shortage of housing. The city has a number of shanty town settlements, such as Rocinha, that house the rural migrants who flock there.
- Rocinha is the largest shanty town in Rio, with a population of around 200 000 people. It is located in the south zone of the city on a steep hillside overlooking the city beaches.
- Not all the people in Rio are poor – many wealthy people live close to the central business district (CBD).

Improving the Quality of Life in Rocinha

- In the 1990s, the Favela Bairro Project (a slum to neighbourhood project) was set up to upgrade the favelas (as opposed to demolishing them) by providing pavements, electricity and sewage systems.
- The project promoted self-help building schemes, where local residents were provided with materials such as concrete blocks and cement to construct permanent dwellings, often with three or four floors and with water and sanitation.
- Residents were given legal rights of ownership or low rents on properties. Improved transport systems gave them better access to work in the city.
- Businesses like shops and restaurants were encouraged.
- Law and order improved through 'pacification' programmes.
- After the area was made safe for visitors, tourism flourished.

Quick Test

1. What percentage of Rio de Janeiro's population live in favelas?
2. Explain three ways in which living conditions in Rocinha favela have been improved.
3. Explain how self-help schemes have improved conditions in Rocinha favela.

Key Point

Rapid urban growth of Rio has presented many social, economic and environmental challenges.

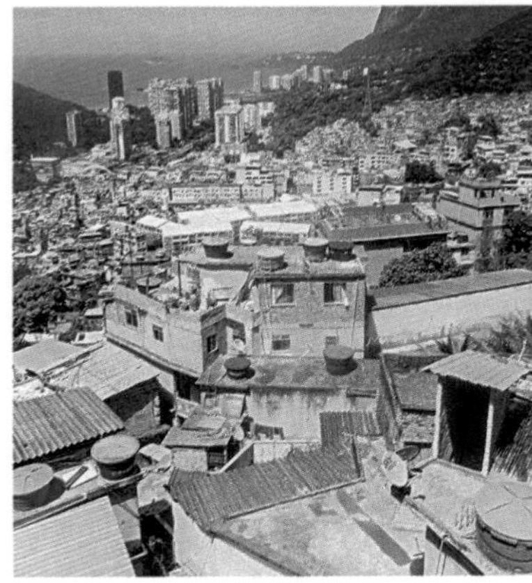

View over Rocinha favela

Key Point

Conditions in favelas can be improved with government support and self-help projects.

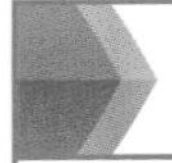

Key Words

favela

Urban Issues and Challenges 2

You must be able to:

- Explain how urban changes in cities in the UK lead to a variety of social, economic and environmental opportunities and challenges
- Explain different ways inner city areas can be regenerated.

Case Study: UK City – London

- Population: 8.7 million (2015).
- Leading global city for industry, education and finance.
- A world cultural capital.
- The most-visited city in the world.
- Diverse range of peoples and cultures.
- Hosted the Olympic Games in 2012.
- During the 19th century, the port of London was the busiest in the world. By 1980, ships had become too large to sail up the River Thames and the docks in the heart of the city became derelict.

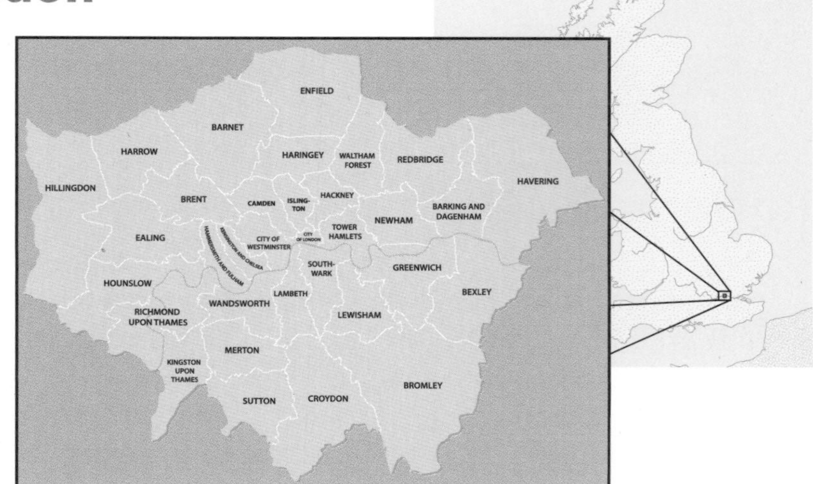

Challenges in London

Social	**Housing inequalities:** House prices in London are greater than anywhere else in the UK. Within the city, there is not enough good quality and affordable housing. Overcrowding has risen (particularly in the private rental sector) and is considerably higher than the rest of the country.
	Education: Overall attainment in London schools matches the rest of the UK, although there is considerable variation across boroughs in the city.
	Health: People with poorer English language skills have found it difficult to access healthcare facilities.
	Urban sprawl: As London has expanded, new buildings have been constructed on greenfield sites around the city.
	Migration: Job opportunities, particularly in finance and 'knowledge-based' industries, have attracted people to London from elsewhere in the UK and the rest of the world. It is estimated that up to one-third of all international migration into the UK is to London.
Economic	**Industry:** As the docks closed, many manufacturing industries were lost.
	Pay inequality: London has the most unequal pay distribution of any part of the UK, largely due to the high wages paid at the top end of the scale. 21% of people living in London are paid below the London living wage.
	Unemployment: The closure of factories led to high unemployment in parts of inner London.
Environmental	**Dereliction:** As manufacturing industries have declined (particularly in the eastern parts of the inner city and along the banks of the River Thames), much land has been left in a state of dereliction.
	Urban sprawl: This has increased pressure on land use in the rural–urban fringe and led to the growth of commuter settlements.
	Waste disposal: A large urban population produces a lot of household and commercial waste for disposal.
	Pollution: Atmospheric pollution from industry and vehicles.

Opportunities in London

Social	**Diversity:** Ethnic and cultural diversity allows people to experience different foods, music and religions.
	Entertainment and culture: London is an international centre for sporting events, theatres, cinemas, museums and art galleries.
Economic	**Industry:** The number of job opportunities in financial services and knowledge-based industries has increased. In addition, the redevelopment of London's docklands and the more recent redevelopment of the Olympic site in East London has increased the number and variety of jobs available in London.
	Transport: A new high-speed train link (HS2) from London to Birmingham and eventually on to Leeds and Manchester is planned. Work continues on the Crossrail project from Paddington station to Reading in the west and Abbey Wood to the east of the city. Plans have been approved to build a new runway to expand Heathrow Airport.
	Tourism: London is the most-visited world city and attracts tourists with its historic buildings, monuments, sports events, cultural events and entertainment.
Environmental	**Reuse of industrial land:** Numerous **brownfield sites** have become available for development.
	Transport: Improvements to transport systems aim to reduce carbon dioxide emissions by 60% by 2025. This includes wider use of diesel-electric hybrid buses, and the trialling of hydrogen fuel cell buses and combined diesel and biofuel buses.
	Regeneration: London's docklands have been regenerated, as part of which the Olympic site has been further developed. Other sites like Battersea Power Station and the Gas Works at Greenwich have also been regenerated for flats. The O2 Arena at Greenwich Peninsula has opened as a multi-purpose indoor arena.

Regeneration of London's Docklands

- The London Dockland Development Corporation (LDDC) spent £10 billion improving this part of London between 1981 and 1988.
- Disused dock basins like the Royal Docks were redeveloped with a mixture of luxury housing, hotels, shops and restaurants.
- New office development took place at Canary Wharf.
- The Docklands Light Railway (DLR) connected the area to the London Underground system.
- London City Airport – a STOL (short take-off and landing) airport.
- Affordable housing for local people.
- New cultural venues like the Docklands Arena and O2 Arena.
- New 'eco' housing, e.g. at Greenwich Millennium Village.
- The London Olympic site left a legacy of housing, sports facilities, improved transport links and the Queen Elizabeth Park.
- The Olympic Stadium is now the home ground for West Ham United.

Key Point

London's docklands fell into decline but have been regenerated with different land uses.

The regenerated Victoria Docks in London's East End area, with the financial district in the background

Quick Test

1. List four modern uses for London's derelict dock areas.
2. Name two planned developments aimed at improving transport in London.
3. What new use was found for the Olympic Stadium after the 2012 Games?

Key Words

brownfield site
regeneration

Urban Issues and Challenges 3

You must be able to:

- Understand why urban areas need to be developed for the future in a sustainable way
- Explain how sustainable living involves a number of changes to current urban lifestyles.

Sustainable Urban Living

- City inputs include food and water, fuels and energy, building materials and consumer goods.
- City outputs include sewage, exhaust gases, household waste, industrial waste and building waste.
- In the UK, 90% of people live in urban areas.
- Cities use huge volumes of resources and produce vast amounts of waste.
- However, urban areas can be developed in a **sustainable** way.
- Sustainable living means allowing people to meet their needs today without harming the prospect of people in the future to meet their needs.
- Sustainable urban living needs to consider ways for people in towns and cities to adapt to climate change and use strategies to mitigate it.

Key Point

Many current trends in urban living are not sustainable. Sustainable urban living includes methods of adaptation to, and mitigation of, climate change.

Features of Sustainable Urban Living

Housing

- Affordable prices and low rents.
- Shared housing.
- Passive housing – kept warm using heat from people, pets and natural lighting.

Resource Management

- The aim is to be **carbon neutral** – for a town to produce as much energy as it consumes.
- Low energy use, e.g. solar heating, efficient insulation, triple glazing, water and electricity meters, low-flow taps, etc.
- Efficient recycling.
- Reduction of household and industrial waste.
- Composting of food and green waste, which counters global warming by reducing methane emissions.
- 'Greywater' and rainwater are collected for domestic use.
- Sedum moss roofing is used to reduce impermeable surfaces, increase lag times and reduce flooding.

Key Point

Sustainable modern cities are self-contained, self-supporting and help to manage climate change.

Urban Transport Strategies to Reduce Traffic Congestion

- Car-sharing schemes.
- Vehicle-restricted areas to reduce congestion and pollution.
- Public transport (including buses, trains and trams) organised into a diverse and efficient integrated rapid transit system.
- Alternative, cleaner fuel sources, e.g. hydrogen and electric. New electric buses planned for Bristol are designed to charge with power when parked over induction plates. These could eventually be installed along the bus routes to allow the vehicles to charge on the move.
- Bus lanes.
- Park-and-ride systems.
- Increased bicycle use, e.g. a system of public cycle hire was introduced in London in 2010 and became known as 'Boris' Bikes', after Boris Johnson, the London Mayor from 2008 to 2016. By 2012, 8000 bicycles had been made available for hire from 570 docking stations distributed around the city.
- Pedestrian walkways and cycle paths, e.g. London's cycle 'super-highways'.

Services and Employment

- Daily needs, such as shops and schools, should be within walking distance for the population.
- There should be a range of job opportunities for residents.

Environment

- 'Urban greening' – a target of 40% green space, parkland and trees.
- Urban forests, e.g. Adelaide in Australia.
- Urban agriculture – vertical farming, allotments and rooftop gardens with the aim of reducing 'food miles'.
- Use of brownfield sites (previously used for industry) for development.
- A green belt of natural space should surround any settlement.

Key Point

Changes to housing and transport can make massive energy savings and help to manage climate change.

Quick Test

1. Identify two ways that modern housing can help to conserve energy.
2. Identify three ways in which modern transport systems can be made more sustainable.
3. What does it mean to describe a sustainable city as self-supporting and self-contained?

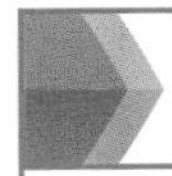

Key Words

sustainable
carbon neutral

Measuring Development and Quality of Life

You must be able to:

- Identify ways by which development and quality of life can be measured and recognise some of the limitations of these measures
- Classify countries according to levels of economic development and quality of life
- Make links between stages of the Demographic Transition Model and level of development.

Measuring Development

- **Development** refers to the progress of a country in terms of economic growth, human welfare and the use of technology.
- **Quality of life** refers to the wide range of human needs that should be met alongside income growth, such as access to housing, education and health, nutrition, security, happiness, etc.
- There is more to development and quality of life than just wealth.
- Geographers use a variety of data (or indicators) to decide how developed a country is. These indicators can cover social, economic or environmental factors such as those shown to the right.
- One single factor does not always tell a full and accurate story about a country. It has been popular in recent times for geographers to use the **Human Development Index** (HDI), which combines life expectancy, literacy and income to measure development.

Indicators used to determine a country's development

- GNI (gross national income) and GDP (gross domestic product), often expressed in US dollars
- Life expectancy (expected number of years of life)
- Literacy (percentage of population that can read and write)
- Birth rate (number of babies born per 1000 people per year)
- Death rate (number of deaths per 1000 people per year)
- People per doctor (population total divided by number of available doctors)
- Calorie intake (total food calorie value per day)
- Car ownership (number of cars owned per 1000 people)
- Access to safe water (measured as a percentage of the total population)
- Infant mortality (number of babies that die before one year of age, measured per 1000 live births each year)

Human Development Index

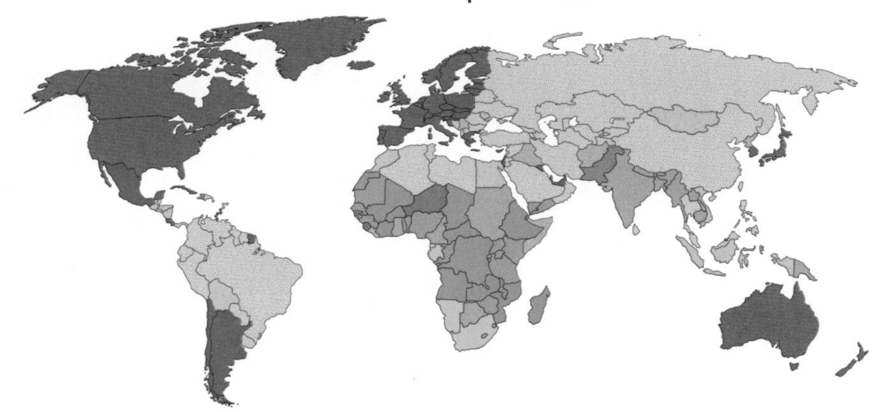

Category	HDI
High human development	0.9–1.00
	0.8–0.89
Medium human development	0.7–0.79
	0.6–0.69
	0.5–0.59
Low human development	0.4–0.49
	0.3–0.39
	0.2–0.29
Not applicable	

Key Point

There are many ways to measure development, apart from wealth.

Classifying Countries by Economic Development and Quality of Life

- Countries can be classified according to their level of development as:
  - **LIC** (lower income country) – classified by the World Bank as less than $1045 GNI (gross national income) per capita, e.g. Ethiopia in Africa.

- HIC (higher income country) – classified by the World Bank as more than $12 746 GNI per capita, e.g. the UK and USA.
- NEE (newly emerging economy) – a country that has begun to experience high rates of economic development, usually with rapid industrialisation. They differ from LICs as they no longer primarily rely on agriculture. The leading NEEs are known as BRICS (Brazil, Russia, India, China and South Africa).

Key Point

Countries develop at different rates and there is a wide range of reasons for this.

Limitations of Economic and Social Measures

- The use of data can give a false picture of development as it gives an average for a whole country. In some countries, such as Brazil and China, there can be great variations in levels of development between regions.
- The use of a single measure of development can be misleading, e.g. low birth rates generally suggest good social development, but birth rates can be reduced by government policy.
- Data can be out-of-date.
- Data can be unreliable due to misreporting or not available due to conflict, disasters or difficulties accessing very remote areas.
- Corrupt governments may distort the available data.

Key Point

Countries with different levels of development and quality of life have different population characteristics.

Development and the Demographic Transition Model

- The Demographic Transition Model (DTM) shows population changes over time. As a country becomes more developed, its population characteristics change.
- The difference between birth rates and death rates represents the natural population increase. If birth rates are higher than death rates, the total population will increase.
- The DTM is divided into five stages. Most HICs are in stage 4 or stage 5, while most LICs are in stage 2 or stage 3.
- Global population is rising rapidly because birth rates are still very high in LICs, while death rates have fallen across the world due to global efforts to tackle malnutrition and diseases such as smallpox.

Demographic Transition Model

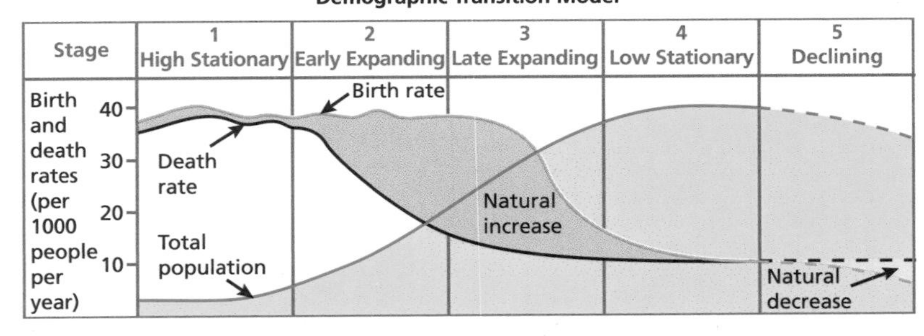

Population Characteristics of HICs

- Growth rate stable
- DTM stage 4
- Low birth rates, low death rates
- Ageing population
- Some countries, like Germany, have declining populations (DTM stage 5)

Population Characteristics of LICs

- Population rising rapidly
- DTM stage 2 or 3
- High birth rates, falling death rates
- Youthful population

Key Words

development
quality of life
Human Development Index
LIC
HIC
NEE
Demographic Transition Model

Quick Test

1. Identify three development indicators (other than wealth).
2. Using the map of Human Development Index on page 76, identify two HICs, two LICs and two NEEs.
3. Who are the BRICS?

The Development Gap

You must be able to:
- Identify reasons why development is uneven
- Explain the consequences of uneven world development
- Examine various strategies to reduce the development gap.

Causes of Uneven Development

Physical
- Extreme climates can affect access to fresh water and the ability to grow food.
- Some countries, such as the Philippines, are affected by high magnitude and/or frequent natural hazards (tropical storms and earthquakes) that hinder development.
- Some countries have limited natural resources, such as fossil fuels and minerals.
- Steep, mountainous terrain is harder to build on and limits farming.
- Some countries are **landlocked**, making international trade difficult. Of the 15 lowest-ranking HDI countries, eight are landlocked (e.g. Chad in Africa).
- In many of the LICs, populations are growing rapidly and putting pressure on resources. This affects access to healthcare, education, water and electricity – all of which are needed for development.

Economic
- Most LICs rely on primary resources for exports. Prices can fluctuate for these products and harvests can fail owing to natural disasters, outbreaks of disease, etc., leaving countries with little income.
- Transnational corporations may exploit LICs, paying low prices for raw materials and food. Sometimes corrupt leaders have sold resources to transnational corporations for personal gain.

Historical
- In the 1700s and 1800s, **colonialism** saw European nations, such as the UK, France and Spain, exploit the resources and labour of many of today's LICs (e.g. Nigeria was a colony of the UK). This meant that countries could not take advantage of their own resources to generate their own development.
- Colonialism ended during the mid-20th century, meaning that many countries in Asia and Africa became independent. Political issues still remain in many of these countries, resulting in war and conflict, which limits development.

Key Point

There is a range of physical, economic and historical reasons for uneven development.

Consequences of Uneven Development

Disparities in Wealth
- The most developed countries have the greatest wealth and a good balance of trade, whereas the quality of life is worsening in some countries.
- Within countries there can be great disparities in wealth, e.g. in Brazil and South Africa.
- LICs have become dependent on HICs for aid, often resulting in debt.

Disparities in Health
- LICs have a shortage of safe, clean water and are unable to invest in good quality healthcare, with infectious diseases such as malaria, tuberculosis and diarrhoea the main cause of deaths – particularly among children.

International Migration

- People seek to move in order to improve their quality of life.
- Some migrants move voluntarily looking for work opportunities, higher wages, better education and healthcare (e.g. many people have moved from countries in eastern Europe to countries in western Europe).
- Some people are forced out of countries owing to conflict, persecution and poverty (e.g. people moving by boat from countries in Africa, like Nigeria, to European countries, like Italy).

Reducing the Development Gap

- 'Development gap' refers to the difference in the standard of living and well-being between the richest and the poorest countries of the world. The gap can be reduced in these ways:

International aid	Can be short-term aid (which solves an immediate problem like lack of homes after an earthquake) or long-term aid (which tries to solve a problem so it never happens again, e.g. building a dam to provide a clean water supply). Aid can be provided by a country, a group of countries or charity groups.
Intermediate technology	Uses simple, appropriate technology that the local community can set up, work with and maintain to improve development (e.g. pumps to provide clean water).
Fair trade	Where producers receive a guaranteed fair price for the things they make and grow. A fair price is one that covers costs of production and enables people to have a reasonable standard of living. It also puts money into community projects like water and schools.
Debt relief	Between 1960 and 1980, many LICs were loaned money from HICs, banks and international organisations. This money was loaned out with high rates of interest and in some cases mis-spent or siphoned off by corrupt governments. Many LICs have been left with debts which they cannot pay back. Cancelling these debts or lowering interest rates can help LICs to develop.
Micro-finance loans	Small amounts of money lent to individuals in LICs, such as by the Grameen Bank, to help them start their own businesses. This enables people to work their way out of poverty and helps development at a local level.
Investment	Many transnational corporations have invested into LICs and NEEs (e.g. Apple manufactures iPhones and iPads in China). This creates jobs and puts money into the local economy.
Industrial development and tourism	Investing in manufacturing and tourism creates greater wealth than primary industries. Governments can tax this wealth, which can then be invested into public services.

Aid can fund pumps to provide clean water

Key Point

Various strategies exist for reducing the global development gap.

Kenya: Using Tourism to Reduce the Development Gap

- Kenya attracts 700 000 visitors each year owing to: wildlife safaris (to see elephants, rhinos, buffalos, lions and leopards); the tribal culture; a warm climate with sunshine all year; varied scenery, including savannah, grassland, mountains, forests, beaches and coral reefs; 23 national parks, e.g. Tsavo and Masai Mara.
- Tourism accounts for 15% of Kenya's GDP.
- Transnational corporations, such as Hilton Hotels, have invested into Kenya.
- Tourism creates jobs directly in hotels and indirectly in supporting industries like taxi companies. There are 250 000 tourism-related jobs.
- Tourists spend money in hotels, restaurants, attractions and local shops, which boosts the economy.
- Infrastructure is improved to cope with the influx of visitors.

Safaris are big business in Kenya

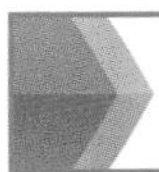
Key Words

landlocked
colonialism
short-term aid
long-term aid
intermediate technology
fair trade

Quick Test

1. Why does international migration occur?
2. Explain the difference between long-term and short-term aid.
3. What is the main principle of fair trade?

Changing Economic World Case Study – Vietnam

You must be able to:

- Describe the changing industrial structure of a particular location and its context
- Examine the role of transnational corporations, political and trading relationships, and international aid
- Explain the impacts of economic development on people and the environment.

Location, Importance and Wider Context

- Vietnam is a country located in south-east Asia (bordering China, Laos and Cambodia). Its capital city is Hanoi. It is a communist country.
- Vietnam is a former French colony. It became independent in 1945.
- Vietnam has a population of 96 million (2018). It is the world's 15th most populated country. The population is growing steadily at a rate of 0.93%. The birth rate is 16 per 1000 and the death rate is 6 per 1000. The country has a youthful age structure.
- 35% of the population live in urban areas, although this is changing with rapid rural-to-urban migration.
- Vietnam has a tropical climate in the south, with hot temperatures and high rainfall all year round. In the north, the climate is monsoonal with a warm rainy season May to September and a dry season October to April.
- Vietnam has a range of natural resources, such as coal, manganese, timber and offshore oil and gas deposits.
- Vietnam has changed from being a low-income country (LIC) to a newly-emerging economy (NEE). Since 2000, it has had one of the fastest rates of economic growth in the world.

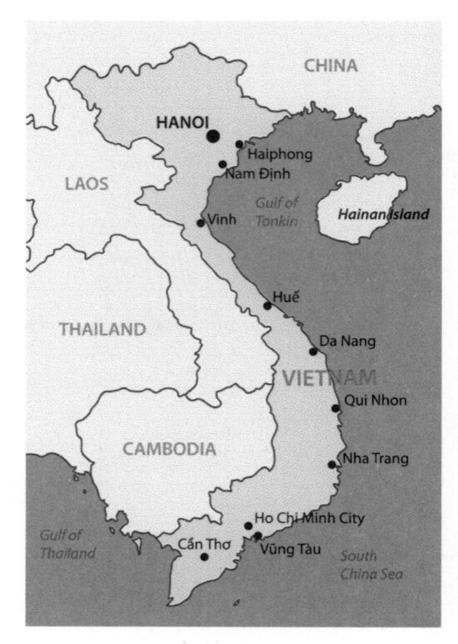

Changing Industrial Structure

- Despite rapid economic growth in Vietnam, 48% of people still work in primary sector jobs such as farming, mining and forestry. The main crops are rice, coffee, tea, soybeans, peppercorns, cashew nuts, fruits and rubber.
- The secondary sector has grown and now employs 21% of people and generates 33% of the country's GDP. This sector is growing rapidly because Vietnam is an important offshore location for **transnational corporations (TNCs)**.
- The service sector is growing rapidly and now employs 31% of people and generates 51% of the country's GDP.
- The tourist industry is growing rapidly in Vietnam with over 4 million international visitors a year.

Role of Transnational Corporations

- Owing to cheap labour, a growing home market and fewer industrial laws and restrictions, TNCs are investing in Vietnam.
- Nike, the sportswear and goods manufacturer, has 34 plants in Vietnam (75% of its workforce is based in south-east Asia). The majority of workers are women under 25.

Advantages of TNC Investment

- Creation of jobs.
- The wages that workers earn are spent in the local economy, creating a positive multiplier effect.
- Investment from foreign companies brings new technology and ideas.
- The government earns tax revenues, which can be invested into infrastructure.

Disadvantages of TNC Investment

- Political influence held over the Vietnamese government as investment can easily be moved away from Vietnam to a lower cost location.
- Many of the jobs created are low paid, especially when compared to jobs in HICs.
- Some profits from TNCs return back to the home country – **economic leakage**.
- New industries can have an environmental impact, including pollution from factories.

Political and Trading Relationships

- Vietnam is becoming economically more important. It has a high annual GDP growth rate of 6.3%, representing its strong manufacturing exports, and rising demand for goods and services within the country.
- By 2050 it is predicted that Vietnam will overtake countries like Portugal, Norway and Singapore in terms of total GDP.
- Vietnam has a trade surplus, meaning that it exports more than it imports.
- Exports include crude oil, clothing, seafood, rice, tea, coffee and electronics. Export destinations include USA, Japan, China, South Korea and Australia.
- Imports include machinery, steel products, raw materials for the clothing and footwear industries, electronics, plastics and cars. The main import partners are China, South Korea, Japan, USA and Thailand.
- Vietnam joined the World Trade Organisation in 2007 and the Association of Southeast Asian Nations (ASEAN) economic community in 2015, showing its growing influence at a regional and global scale.

Key Point

Countries like Vietnam are newly-emerging economies and see manufacturing industry and the export trade as ways to help reduce the development gap.

International Aid

- Vietnam receives multilateral aid from the World Bank. This is to help sustainable development and to promote good governance.
- Bilateral aid is given to Vietnam from countries such as Australia and the USA. In 2017–2018 Australia was to provide $484.2 million for projects such as improving transport infrastructure, training workers and promoting women's economic empowerment.
- Voluntary aid is being given to Vietnam by charities, e.g. Action Aid has been working on women's rights, tackling hunger and providing education.

Effects on Environment and Quality of Life

- Industrialisation and urbanisation (e.g. in Hanoi and Ho Chi Minh City) are causing rising levels of air and water pollution, and waste.
- Logging and farming are causing deforestation with resulting loss of animal habitats, increases in flood risk and soil erosion.
- Overfishing is threatening marine life populations.
- Energy consumption has risen with economic development. 58% of electricity is generated by fossil fuels, which release greenhouse gases.
- Growth in the economy has seen people's incomes rise significantly.
- Poverty levels have decreased from 58% in 1993 to 3.2% in 2016.
- Having more money means people can improve their quality of life, e.g. access to clean water, healthcare and better housing.
- There are significant inequalities in quality of life, with people in rural areas and from ethnic minorities more likely to live in poverty.
- It has been reported that some people (e.g. those employed by TNCs) have to work long hours in difficult conditions.

Forest cleared for agriculture in the Vietnamese highlands

Quick Test

1. Identify three reasons why a TNC might want to locate in a country like Vietnam.
2. What is the difference between multilateral and bilateral aid?
3. List three negative environmental impacts resulting from economic development in Vietnam.

Key Words

transnational corporations (TNCs)
economic leakage

UK Economic Change

You must be able to:

- Understand the causes of economic change in the UK
- Explain what is meant by the North–South divide and be aware of strategies that have been used to resolve regional differences
- Discuss the social and economic changes in rural landscapes; in an area with population growth and in an area with population decline.

The Changing UK Economy

- The economic structure of the UK has changed over time.
- Since the 1980s, there has been a decrease in primary and secondary employment and this is known as **deindustrialisation**.
- Primary employment, such as farming and mining, has declined due to increased mechanisation and automation, meaning that fewer workers are required, and the depletion of locally available raw materials.
- Secondary employment, such as the manufacturing of clothes, has declined owing to competition from overseas where labour costs are cheaper and production methods are more advanced.
- Government policy, such as investment in infrastructure like roads and industrial estates, financial incentives and faster planning permission, has attracted some industry to the UK (e.g. Honda, the Japanese car manufacturer, opened a car plant near Swindon).
- Since 2001, there has been an increase in tertiary employment, e.g. tourism, financial services and healthcare.
- Since 2001, there has also been an increase in quaternary employment – 'knowledge-based' industries such as software companies, research and development, and biotechnology in science and business parks on the edges of cities like Bristol and Cambridge.
- In recent years there has been an increase in flexible part-time working, job sharing and home teleworking using the Internet.
- The economic structure varies around the UK, with different areas specialising in different industries (e.g. London is a financial centre).

Positive Impacts of Globalisation on the UK Economy

- Cheaper goods and services as they are produced in places where labour costs are lower.
- Investment from foreign companies bringing new technology and ideas.
- Migrants fill jobs where there is a shortage of workers, e.g. agriculture and construction.

Negative Impacts of Globalisation on the UK Economy

- Business closures and loss of jobs as companies close down and move overseas, where labour costs are cheaper.
- Loss of skilled British workers who migrate overseas for higher wages.
- Greater pay inequality between highly skilled and unskilled workers.

UK Economic Structure 2001

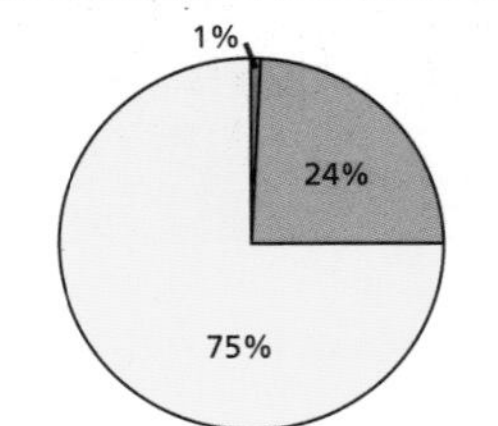

UK Economic Structure 2015

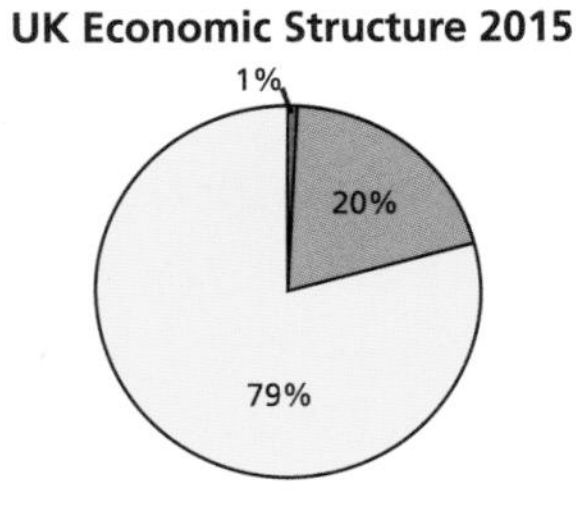

- Primary industry (extractive industrie
- Secondary (manufacturing)
- Tertiary (services) and quaternary (IT and Research and Development)

Key Point

There has been a shift away from traditional forms of employment in recent years.

The North–South Divide

- The UK demonstrates many regional patterns.
- The North has depended on heavy industries like steel-making and ship-building, which have suffered a decline in recent years, while London and the South-East have developed rapidly owing to a fast-growing service sector, e.g. finance, high technology, research and media.

- Wages are higher in the South. Unemployment is higher in the North.
- Average house prices are higher in the South owing to higher demand.
- Population is growing more rapidly in the South owing to more work opportunities and higher wages.
- There are exceptions to the North–South divide and there are large inequalities within particular areas, such as in London.

Strategies to Resolve Regional Differences

- The Government has created the 'Great Northern Powerhouse' initiative.
- The Government has agreed devolution measures to give extra power and money to directly-elected mayors to ensure decisions affecting the North are made by the North.
- Improvements to transport infrastructure linking cities in the North, making it a more attractive place to live and work. Manchester and Leeds will be connected to HS2, a high-speed rail link between the UK's major cities that is due to open in 2026.
- The Government has established **enterprise zones**, providing tax breaks and government support. They have attracted new and expanding businesses, creating jobs and supporting economic growth.

Social and Economic Change in Rural Areas

An Area with Population Growth

- **Counterurbanisation** is where people move from urban areas to surrounding rural areas for a better quality of life. They then typically commute to work in a nearby urban area.
- Bramhall is a village 10 miles south of Manchester. It is a popular place in which to live, having a range of housing, services and good transport connections. Bramhall has its own railway station, is near to motorways such as the M56 and M60, and is close to Manchester Airport.
- In 1911, the population of Bramhall was approximately 2500; by 2011 it had increased to around 17 500.

An Area with Population Decline

- The Outer Hebrides is a set of islands off north-west Scotland.
- It is an extremely rural area with a population density of 9 people per km^2.
- In 1911, the population of this area was approximately 45 000; by 2011 it had gradually fallen to about 30 000.
- The economy of the area is largely based on farming and fishing, meaning there are few job opportunities. There is limited transport availability and housing opportunities.
- Young people have moved away to find job opportunities, leaving behind an ageing population.
- Owing to population decline, there is less demand for shops and services, resulting in business closures. Fewer people also mean that the local council has less money from taxes to invest in public services like healthcare and schools. These services get reduced and close. The area is not attractive to inward investment, so buildings are left derelict. As the quality of life deteriorates, more people move away; this is known as a **negative multiplier effect**.

Quick Test

1. Identify three reasons for deindustrialisation in the UK.
2. What is an 'enterprise zone'?
3. List two economic and two social impacts of population growth in a rural landscape.

Positive Impacts of Population Growth on Bramhall

- Population growth has supported local services and led to new shops and services opening.
- New local businesses have started up.
- Old housing has been modernised.
- There is a good community spirit with lots of community events throughout the year.

Negative Impacts of Population Growth on Bramhall

- House prices are pushed up, forcing some people out of the area.
- Services become orientated towards wealthy people rather than meeting the needs of the local community.
- Local schools are oversubscribed.
- At commuting times there is heavy congestion on local roads, causing noise and air pollution.

Key Point

Population change has caused major economic and social changes in rural areas.

Key Words

deindustrialisation
enterprise zones
counterurbanisation
negative multiplier effect

UK Economic Development

You must be able to:

- Describe different types of links between the UK and the wider world
- Outline improvements and new developments in transport
- Explain how industry impacts on the environment and suggest ways in which industrial development can be made more sustainable.

The Place of the UK in the Wider World

- **Globalisation** is the process by which the world is becoming increasingly interconnected in terms of industry, markets, migration and cultures. The UK has strong links with the wider world.
- The UK trades with countries around the world, particularly those in Europe which are geographically close and economically wealthy. The UK also trades with countries such as the USA and China.
- Investment by individuals and firms from abroad (known as foreign direct investment) has made a major contribution to the UK economy. For example, Honda, the Japanese car manufacturer, has a plant in Swindon.
- The UK has strong creative industries – music, television, film, etc. These products are exported worldwide.
- The Channel Tunnel links the UK to France and therefore mainland Europe. There are a number of international airports in the UK. Transport links enable people and goods to move between countries.
- People and businesses in the UK have very good access to telecommunication and Internet networks, enabling communication, marketing and financial transactions to take place at a global scale.
- The UK has been part of the European Union, a group of 28 countries that trade together favorably and allow freedom of movement of people. After voting to leave the European Union in 2016, the UK started renegotiating relationships with other countries.
- The UK also works closely with Commonwealth countries, an association of 53 independent states, many of which were part of the former British Empire. They share values of democracy, human rights and the rule of law. The UK is linked to many of these countries through trade, culture and migration.
- The UK is part of the G7 (the seven most powerful industrialised countries) with the USA, Canada, France, Germany, Italy and Japan. These countries are very important in global decision-making.

Key Point

The UK is a major player in the global economy.

UK Transport Developments

- **Infrastructure** improvements in the UK include road building and improvement programmes. In some places, motorways are being widened with an extra lane. Many motorways (e.g. the M6) are being made 'smart', whereby speed limits are varied to try to keep traffic moving and to reduce congestion.
- There has been government investment in railways. A high-speed rail link runs from London to the Channel Tunnel. Crossrail 1 and 2 are being constructed to improve transport in the London area.
- The HS2 link should reduce pressure on roads and rail networks and will significantly cut journey times between major cities.

A high-speed train operating between London and the Channel Tunnel

- UK airports are operating at near-capacity. There are proposals to expand London Heathrow, the UK's largest airport, by building a third runway. There has also been talk of building a second runway at Manchester Airport, helping to boost the economy in northern England.
- A new port on the Thames estuary, called London Gateway, opened in 2013. It can accommodate large container ships and enables goods to be brought close to London, the largest consumer market in the UK.

Impacts of Industry on the Environment

- All types of industry can have negative effects on the environment.
- Primary industries have a huge impact on the environment. Farmers use herbicides, pesticides and fertilisers, which can run off into water sources and impact supplies and ecosystems. Mining and quarrying create lots of dust and noise pollution and can leave permanent scars on the landscape.
- Manufacturing industries construct large factory buildings that can destroy **greenfield sites** and intrude on the landscape.
- Tertiary and quaternary industries create lots of waste, which can pollute the landscape and the air through waste-disposal processes.
- Most industries use lots of electricity, which is often created by burning fossil fuels. This creates greenhouse gas emissions, contributing towards the enhanced greenhouse effect and therefore climate change.
- Transport of raw materials and finished products for different industries (usually by road) increases noise and air pollution.
- Traditional UK industries often caused a lot of pollution. Modern industry is being made more environmentally sustainable through:
  - use of modern technology (like desulphurisation) to reduce emissions from factories and power stations;
  - use of renewable sources of energy;
  - investment into energy efficiency;
  - stricter environmental targets for water quality, pollution and landscape damage;
  - heavy fines for industrial pollution;
  - choosing **brownfield sites** for development.
- Adnams is a brewery producing beers and spirits based in Southwold in Suffolk. Since 2017, the business has switched entirely to renewable sources of electricity – wind, solar and hydro. This has massively reduced the company's carbon emissions and therefore its contribution to the enhanced greenhouse effect and climate change.
- Adnams is trying to reduce its water consumption, which is very important given that Southwold is located in the UK's driest region. The business is working hard to reduce packaging. It has reduced the amount of glass used in beer bottles, and it charges customers for carrier bags and donates the proceeds to the Suffolk Wildlife Trust. The company tries to ensure that buildings blend in with the landscape as best as possible and try to ensure that biodiversity is preserved.

Road transport is a source of pollution

Key Point

Industrial development can lead to a wide range of negative impacts on the environment.

Adnams Brewery Distribution Centre with a green roof, solar panels and lime/hemp walls

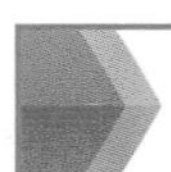

Key Words

globalisation
infrastructure
greenfield site
brownfield site

Quick Test

1. What is meant by the term 'globalisation'?
2. List two benefits of the Government's investment into high-speed railway networks.

Review Questions

Glaciation 1: Processes

1. Where is supra-glacial moraine? [1]
2. The diagram below shows a glacier transporting material.

   What types of moraine are represented by A, B and C?

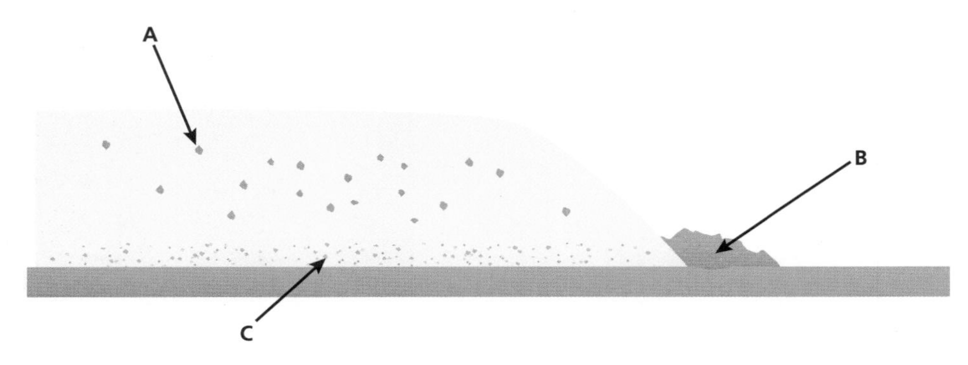

[3]

3. Why do glaciers deposit material? [4]

Total Marks .... / 8

Glaciation 2: Landscape

1. Where would you expect to see a ribbon lake? [1]
2. Why would a valley with lots of tributaries become a very deep glaciated valley? [2]
3. Describe how abrasion shapes a corrie. [3]

Total Marks .... / 6

Glaciation 3: Land Use and Issues

1. Describe the climate of a named upland glaciated area in the UK. [4]
2. How can farming in glaciated uplands damage the environment? [4]

Total Marks .... / 8

Coasts 1: Processes

1. What is meant by a 'constructive wave'? [2]
2. How might a beach change if there are a lot of surging waves? [2]
3. Describe hydraulic action and the results. [3]

Total Marks .... / 7

Coasts 2: Landforms

1. How are wave cut platforms formed? [3]
2. Explain why headlands are eroded but bays become places of deposition. [4]

Total Marks .... / 7

Coasts 3: Management

1. Why is a beach a good form of coastal protection? [3]
2. Why are sand dunes easily damaged by humans? [3]
3. Using the diagrams below, describe some of the changes resulting from building at the coast.

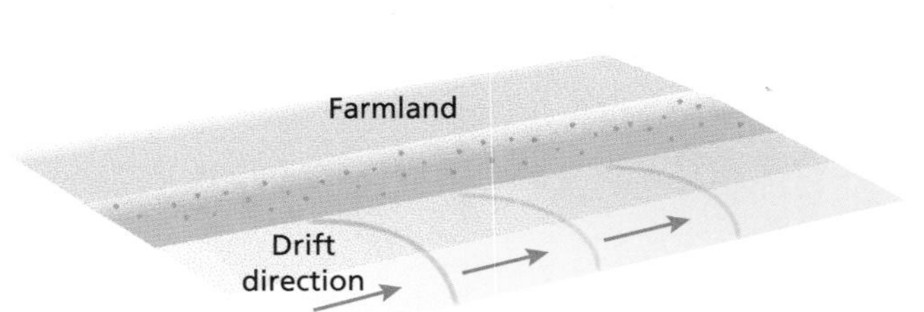

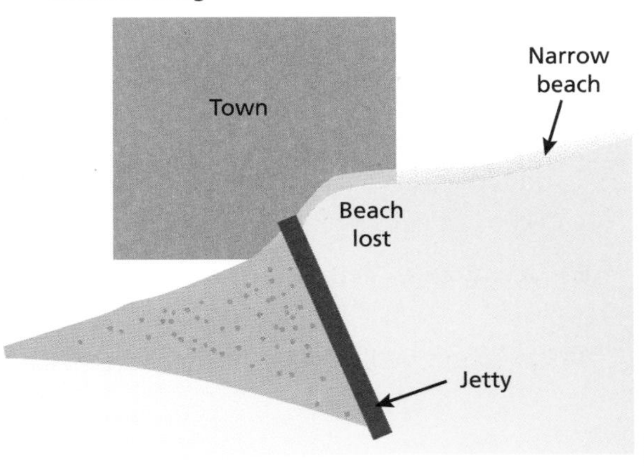

[4]

Total Marks .... / 10

Review Questions

Rivers 1: Processes

1. Explain why a particle of sand is easier to entrain than a particle of silt or clay. [2]
2. Describe what could happen to the river bank in the diagram as a result of abrasion. [3]

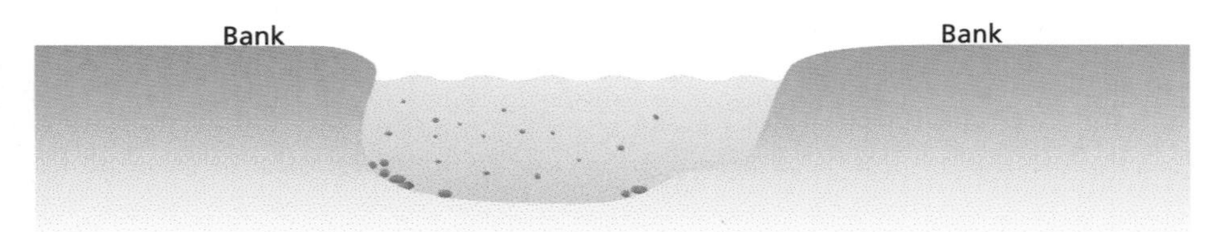

3. Explain two reasons for deposition taking place in rivers. [4]

Total Marks .... / 9

Rivers 2: Landforms

1. Why do spurs often 'interlock' in upland and small rivers? [3]
2. On the diagram below, indicate the location where a waterfall is likely to form. [1]

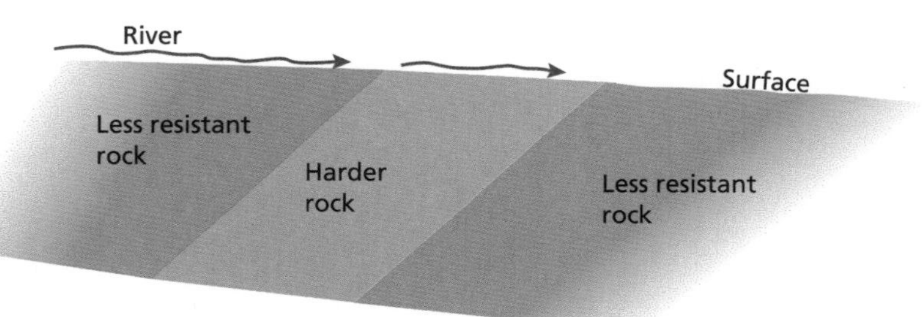

Total Marks .... / 4

Rivers 3: Flooding and Management

1. What is a storm hydrograph? [2]
2. What effect will a very high water table have on the ability of a channel to hold water? [3]
3. Referring to a specific place or flood event, explain **one** human influence on flooding. [3]
4. Describe two **natural** reasons why a specific place flooded / might flood. [4]

Total Marks .... / 12

Practice Questions

Urbanisation

1. What percentage of the world's population lived, or is predicted to live, in urban areas in the following years? Choose from the following possible answers: **10% 25% 42% 54% 60%**

   a) 1900 [1] **b)** 2015 [1] **c)** 2030 [1]

2. What does the term 'urban sprawl' mean, and why might it be considered a problem? [3]

3. **a)** List some of the different push factors that lead to rural to urban migration. [3]

   b) List some of the different pull factors that lead to rural to urban migration. [3]

4. Explain the differences in urban growth between the richer and poorer parts of the world. [4]

Total Marks .... / 16

Urban Issues and Challenges 1

1. Explain why so many rural people are attracted to cities like Rio de Janeiro in Brazil. [4]

2. Rio de Janeiro has a large number of shanty towns. What problems are faced by the people who live in them? [4]

3. Tourism is a major industry in Rio de Janeiro. Describe some of the attractions the city has for visitors. [4]

4. Describe in detail two social challenges of life in Rio de Janeiro. [4]

Total Marks .... / 16

Urban Issues and Challenges 2

1. Why did the docklands in east London fall into decline in the 20th century? [2]

2. Identify four social improvements that formed part of the legacy of the London Olympics. [4]

3. Explain two ways in which transport systems can be improved in London to help reduce carbon emissions. [2]

4. Describe two new projects that aim to improve transport networks in London. [4]

Total Marks .... / 12

Practice Questions

Urban Issues and Challenges 3

1. What percentage of the UK population live in urban areas? [1]
2. Cities can sometimes be described as systems. Name two inputs into a city system. [2]
3. Describe in detail three different strategies that can help to reduce traffic congestion. [6]
4. What changes can be made to modern cities to improve resource management? [6]

Total Marks .... / 15

Measuring Development and Quality of Life

1. Name two social indicators that could be used to measure development. [2]
2. Name two economic indicators that could be used to measure development. [2]
3. Identify four population characteristics of a HIC. [4]

Total Marks .... / 8

The Development Gap

1. Describe two physical causes of global inequalities. [4]
2. How can 'fair trade' help to improve life for farmers in LICs? [4]
3. What attractions can Kenya offer to bring in tourists? [4]
4. How can tourism help to reduce the development gap in Kenya? [4]

Total Marks .... / 16

Changing Economic World Case Study – Vietnam

1. What factors attract many TNCs to locate in NEEs? [3]
2. Explain what is meant by 'economic leakage'. [1]
3. Describe the industrial structure of a NEE you have studied. [3]

Total Marks .... / 7

UK Economic Change

1. Name two cities where science parks have developed. [2]
2. Complete the following table of employment data in the UK:

2001	2015 – higher, same or lower?
Primary = 1%	
Secondary = 24%	
Tertiary and Quaternary = 75%	

[3]

3. Outline strategies that have been used to reduce the north–south divide in the UK. [3]

Total Marks .... / 8

UK Economic Development

1. List two of the UK's main trading partners. [2]
2. Explain how industrial development can be made more sustainable. [3]
3. Distinguish between a greenfield site and a brownfield site. [2]

Total Marks .... / 7

Overview of Resources – UK

You must be able to:

- Understand how importing food products from abroad can increase carbon footprints and food miles
- Explain how demand for water is increasing in the UK
- Appreciate how the sources of energy used to power the UK are changing.

Food in the UK

- Only around 25% of all the fruit and vegetables consumed in the UK are grown there.
- British supermarkets sell fruit and vegetables like green beans, mangetout, peas and strawberries that are imported from abroad, especially from countries that have climates allowing these products to be grown all year.
- Most **organic** produce sold in UK supermarkets is grown within the UK but is only sold when 'in season'.
- The import of fruit and vegetables from countries like Kenya and Peru leads to increased **food miles** and **carbon footprints** but does help the economies of those countries.
- There is an increasingly popular trend to only eat locally produced meat, fruit and vegetables only when in season, thereby cutting down on food miles.
- In the UK, agribusiness has become more important in the last 40 years as farm and field sizes have increased along with the use of machinery and agrichemicals, while the number of farm workers has decreased.

Key Point

The UK is reliant on imported food from around the world; a process that increases its carbon footprint.

Water Use in the UK

- With increasing affluence, water demand is growing as new houses are built with more than one bathroom and consumers demand labour-saving devices such as dishwashers.
- Industry and agriculture are also large users of water, e.g. 15000 litres of water are needed to produce 1 kg of beef.
- Maintaining the quality of drinking water and the management of water pollution are costly, requiring great investment in technology and with costs that are largely met by the consumer.
- The areas of highest demand in the UK – London and the South East – are the areas where there is least available water.
- The areas with least demand for water – the north and the west of the UK – are also the wettest and have surplus water.
- To maintain water supplies, water is transferred from wetter regions to those drier areas with greatest demand. For example, water is sent from Kielder Water in Northumberland to cities like Leeds using a system of pipelines, aqueducts and rivers.

Key Point

Increasing water consumption is leading to water supply problems in some parts of the UK.

Energy Use in the UK

- In the 1960s, most of the UK's electricity was generated using coal, oil and a small amount of nuclear (gas was also made from coal).
- Today, around half of the UK's electricity is made from gas and the rest is from coal, nuclear and **renewables**.
- The reliance on fossil fuels has a significant impact on the UK's energy security. Increasing amounts of energy are being imported because domestic supplies of coal, gas and oil are running out.
- Some see nuclear power as a way of reducing the UK's dependence on fossil fuels as it is a low carbon alternative, but it has many disadvantages.
- The use of renewables is growing in significance as the number of wind and solar power installations increases.
- The development of renewable energy is expensive and the cost to the consumer is higher than electricity from fossils fuels.
- New methods of exploiting energy sources include shale gas that is obtained through **fracking**, a hydraulic fracturing process that has many environmental issues associated with it.

Key Point

Alternative sources of energy will become increasingly important in the future.

Fracking

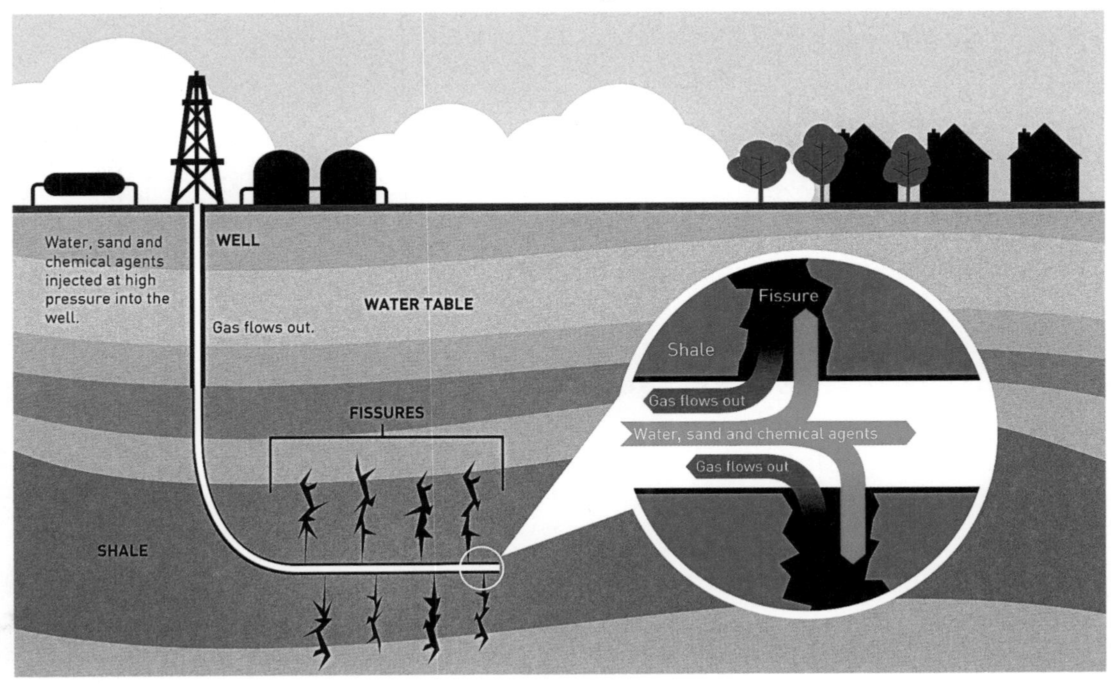

Quick Test

1. Why does the UK need to import food from abroad?
2. Why is water demand increasing in the UK?
3. Why do London and the South East have problems in supplying enough water?
4. What challenges does the UK face to provide enough energy in the future?

Key Words

organic
food miles
carbon footprint
renewables
fracking

Food 1

You must be able to:

- Understand that the global demand for food is rising but supply can be insecure, which may lead to conflict
- Explore the reasons for increasing food consumption and factors affecting food supply.

Food Security

- In many developed countries, **food security** is the norm and the population are well-fed and nourished.
- More than 800 million people, mainly in developing countries throughout the world, live with hunger or **food insecurity**.
- Food production is influenced by two factors – the physical environment and human capacity – and is not evenly distributed throughout the world.
- Climate, water availability and soil type are the main components of the physical environment, while human capacity refers to population size, the farming skills and the financial investment a country is able to put into its agriculture.
- Countries with vast populations like China and India have agricultural output values in excess of $100 billion.
- HICs such as the USA have very large output values due to the intensification of farming methods and large investments of capital.
- In contrast LICs, like those found in Sub-Saharan Africa, produce less food.
- Calorie intake also varies between rich and poor countries.
- People in the USA and EU consume the most calories per capita on average (3000 per day).
- This is much higher than the recommended daily calorie intake of 2500 for men and 2000 for women, leading to obesity in these countries.
- Sub-Saharan Africa has many countries experiencing below average per capita calorie intake.

Poor physical conditions can contribute to food insecurity

Increasing Food Consumption

- Many LICs have rising populations and, as the world's population increases by around 80 million people each year, more land is put under cultivation to feed them.
- As countries develop economically, their population demands a more western-style diet with more meat and dairy products.
- This leads to grain that was once used to feed the people being fed to animals, resulting in shortages for the poor and diet-related illness for the rich.

Key Point

HICs have high levels of food security while LICs tend to have low levels.

Factors Affecting Food Supply

- Extreme climatic conditions (such as drought or floods) can lead to low agricultural yields and environmental degradation (such as desertification), in turn leading to famine.
- Famine is more complex than just a shortage of food. The UN definition of famine is complex and has three conditions:
  - More than 20% of the population must have fewer than 2100 calories of food available per day.
  - More than 30% of children must be acutely malnourished.
  - At least two deaths per day in every 10 000 people or four deaths per day in every 10 000 children must be being caused by lack of food.
- Famine causes rising food prices. Other factors can also cause rises to food prices, such as natural disasters like drought, leading to under-nutrition and starvation among the poorest in those countries.
- **Climate change** has also had an impact in some of the countries of Africa, as soil erosion and desertification have led to a lack of land to grow food on.
- The countries that face the biggest food shortages, and therefore the worst levels of food insecurity, are often those that face the biggest threats from agricultural pests and diseases.
- Those countries also face the biggest risk from water shortages caused by unpredictable rainfall.
- The Sahel region of Africa has been suffering from drought on a regular basis since the early 1980s. The 1983–85 famine in northern Ethiopia was the worst to hit the country in a century and led to 400 000 deaths.
- The effects of the famine were made worse by a background of social unrest and civil war, which had raged in Ethiopia and Eritrea for two decades.
- The introduction of technology into agricultural methods in LICs can lead to massive growth in the amount of food being produced.

Poor soil conditions hinder food production

Key Point

Producing food sustainably is a challenge for most nations.

Technology can help to increase food production

Quick Test

1. What is food security?
2. Suggest physical reasons for food insecurity.
3. Suggest human reasons for food insecurity.

Key Words

food security
food insecurity
yield
famine

Food 2

You must be able to:

- Describe the range of strategies that increase food supply
- Explain the ways in which food production can be made more sustainable.

Increasing Food Supply

- Irrigation is the application of water to increase the yields of crops, especially in those areas where water is in short supply.
- It is also used in HICs where the plentiful supply of water can be seasonal.
- The Gezira Scheme in Sudan is an example of a large-scale irrigation project, where the floodwaters of the Nile are used to grow crops such as cotton and wheat. However, large amounts of water are taken out for this process, which affects other irrigation schemes downstream.
- Hydroponics is a method of growing plants using nutrient-rich water, often using inputs of ultra-violet light; for example, growing high value crops like tomatoes in greenhouses.
- Aeroponics is a similar way of growing plants in an air or mist environment, without the use of soil.
- The Blue Revolution refers to the growth of aquaculture as a very highly productive way of producing food, such as aquatic animals and plants in both saltwater and freshwater.
- Fish farming, such as shrimp or salmon farming, and the gathering of seaweed are all methods of aquaculture.
- The Green Revolution is an initiative that has increased agricultural production, particularly in LICs, since the late 1960s.
- The Green Revolution saved around a billion people from starvation and involved the development of high-yielding varieties of cereals (rice and wheat) and new farming techniques, including the use of irrigation, synthetic fertilisers and pesticides.
- Biotechnology uses biological processes to develop new products that can help feed the hungry by producing higher crop yields.
- Biotechnology can lower the input of agricultural chemicals into crops, resulting in fewer vitamin and nutrient deficiencies and reduced allergens and toxins.
- Appropriate technology is small-scale technology that is simple enough that people can manage it directly and on a local level.
- It makes use of skills and technology that are available in a local community to supply basic human needs, such as gas and electricity, water, food, and disposal of waste. Examples include the Jiko Stove, which is made from locally sourced materials.

Irrigation

Key Point

Technology can be used to increase food supply and reduce food insecurity.

Jiko Stove

Making Agriculture More Sustainable

- Organic farming is a sustainable method of food production that does not use synthetic pesticides or fertilisers. Organic products, such as plant and animal waste, are used as fertiliser and biological solutions have been developed for pest and disease control.
- Urban farming is a sustainable practice where growers cultivate and process produce, often for their own use, and can include a variety of activities such as vegetable and fruit growing or beekeeping.
- In LICs, where poverty is a problem, urban farming is often practised by urban dwellers to produce enough food for their families.
- The Organopónicos in Havana, Cuba, is a good example of a local scheme to increase food security for the urban poor.
- The scheme started in the early 1990s when the Soviet Union collapsed and could no longer provide food to Cuba. The Organopónicos consist of low-level concrete walls filled with organic matter and soil, with lines of drip irrigation laid on the surface.
- Around 35 000 hectares of Organopónicos are found in Havana, growing various fruit and vegetables.
- In HICs, allotments and smallholdings are popular leisure pastimes as well as producing plentiful and cheap, nutritious food.
- **Ethical consumerism** is the purchase of sustainable goods, such as sourcing food that is locally grown and only eating fruit and vegetables that are in season.
- Other examples include changing diets to those less reliant on meat and dairy products, or using sustainable fish sources such as pollock instead of cod (ensuring that the cod will recover in overfished areas).
- Reduced food waste and losses in both homes and in shops could save UK consumers up to £2.4 billion a year.
- The average UK family throws away enough food for around six meals a week, so buying less food in the supermarket and using goods before they go 'off' would save the average family around £60 a month.
- **Permaculture** is a method of agriculture where humans work with nature to farm sustainably.
- Sustainable management of the tropical rainforest often involves permaculture to harvest products such as wild game, nuts and berries.

Key Point

Sustainable use of food resources can also lead to increased food supply and reduced food insecurity.

Throwing away food is a waste of money

Quick Test

1. How does technology increase crop yields?
2. How can changing our eating habits ensure that we can reduce our carbon footprint?
3. What is ethical consumerism?

Key Words

irrigation
ethical consumerism
permaculture

Water 1

You must be able to:

- Explain why some areas have water surpluses and some have deficits
- Describe how and why water use varies in different countries
- Consider the impacts of water insecurity.

Water Supply

- Demand for water resources is rising globally but supply can be insecure, which may lead to conflict.
- The supply of water is uneven because the distribution of rainfall varies from place to place.
- Regions of the world that have water surplus have water security, while those with a deficit have water insecurity.
- For example, the UK, France and Germany have water security; Turkey, Israel and Jordan have water insecurity.
- Some regions have a physical water scarcity through natural reasons for low water supply, e.g. low rainfall in deserts.
- Economic water scarcity is low supply caused by human reasons, such as the lack of a water infrastructure that can be found in many LICs.
- The wettest places in the world, such as rainforests, do not support many people, while the number of people living in arid regions is increasing and putting a greater demand on available resources.
- As the world's population rises so does the demand for water for drinking, bathing, agriculture and industry.

Water Usage

- With economic development, the demand for water rises because domestic use increases through more labour-saving devices like washing machines and dishwashers.
- The average North American uses 570 litres of water per day, the average European uses 130 litres, and the average African around 20 litres.
- With economic development, industrial and agricultural use of water also increases due to a greater demand for goods and services, such as meat and dairy products.
- Regions that experience low rainfall will often have limited groundwater and an absence of surface water supplies, which in turn can lead to reduced water availability.
- Certain types of rock (such as limestones and sandstones) store water. Other types of rock (such as clays and granite) only allow for the storage of water on the surface.

Key Point

Water use increases with rising levels of development and population growth.

- Increasing human water usage has resulted in groundwater supplies – a non-renewable resource – being depleted, while rivers are often diverted for agricultural and industrial purposes.
- In many LICs, regions with limited water supplies lose up to 75% of water used in irrigation through leakage, evaporation or run-off.
- Fresh water is very rare – only 3% of the world's water is fresh, with two-thirds of that stored in ice caps and glaciers and unavailable for human use.
- Agriculture consumes more water than any other user and wastes much of that through inefficient management.
- It also accounts for one of the most common types of water pollution, from agricultural chemicals.
- Sources of water pollution also come from factories, vehicle emissions and human effluent.

Key Point

Agriculture is the biggest user of water.

Water in LICs

- Some 1.1 billion people worldwide lack access to safe drinking water, and a total of 2.7 billion find water scarce for at least one month of the year.
- Poor sanitation is a problem for 2.4 billion people as they are exposed to diseases, such as cholera, typhoid and other **waterborne diseases**.
- Two million people, mostly children, die each year from diarrhoea.
- Many rivers, lakes and aquifers are drying up or becoming too polluted to use and more than half of the world's wetlands have disappeared.
- Climate change is altering patterns of rainfall around the world, causing shortages and droughts in some areas and floods in others.
- This has serious implications for agriculture, which could lead to food shortages and famine in areas of the world like the Sahel region of Africa.
- Over-extraction of water for energy production and industrial processes can lead to shortages and a possible source of conflict between countries.
- By 2025, two-thirds of the world's population may face water shortages.

Key Point

Over 1 billion people do not have access to safe drinking water.

Quick Test

1. Define 'water security'.
2. Why are water supplies in some places becoming scarce?
3. Why is access to clean water difficult in some LICs?

Key Words

water security
physical water scarcity
economic water scarcity
arid region
groundwater
waterborne diseases

Water 2

You must be able to:

- Describe schemes to transfer water between regions with surpluses and those with deficits
- Explain strategies to obtain and conserve water sustainably.

Increasing Water Supply

- In countries were water is scarce, diverting supplies from one river system to another is widespread.
- The building of dams and reservoirs also increases the supply of water for multiple uses by the population.
- **Water transfer schemes** such as the South–North Water Transfer Project in China will enable 44.8 billion cubic metres of water per year to be transferred from the Yangtze River in southern China to the Yellow River Basin in arid northern China.
- The cost of the project is over $60 billion, but it is likely to cause water shortages in some parts of China and pollution in some rivers.
- As climate change makes droughts more common, a growing number of countries are turning to **desalination** (removing salt from either seawater or groundwater to make it drinkable).
- Even London now has a seawater desalination plant as the South East is a water-stressed area, especially during dry years.

Key Point

In countries where there is a water shortage, water is transferred from regions with surpluses to those with deficits.

Sustainable Water Conservation

- Water can be conserved through sustainable management strategies.
- **Water conservation** methods can take place in the home and garden and can include low-flush toilets, taking short showers and only using washing machines and dishwashers when full.
- In the garden, methods can include growing drought-resistant plants that need little watering, spreading mulch or composted bark on flower borders and connecting a water butt to drain pipes and using the rainwater to water plants.
- Groundwater makes up nearly 30% of all the world's freshwater and is often used in places where there is not enough water to drink.
- Groundwater is often found in porous rocks deep underground in reservoirs called **aquifers**.
- Groundwater quality is usually very good and needs less treatment than river water to make it safe to drink, as the rocks through which the groundwater flows help to remove pollutants.

A water butt

- Groundwater also responds slowly to changes in rainfall, and so it stays available during the summer and during droughts, when rivers and streams have dried up.
- Groundwater is relied on in many developing countries, especially those in Africa, because it can often be found close to villages.
- Extracting groundwater is relatively inexpensive as it is often drawn from wells – this does not require much technology and means that reservoirs are not needed to store the water before it is used.
- Aquifers are often slow to recharge (fill up) so may not always be sustainable.
- Water used in the home can be recycled by treating it in waste water (or sewage) plants.
- **Greywater** harvesting is a way of conserving water, involving the recycling of water used in baths and showers, as well as rainwater from roofs, to be used for flushing toilets and other non-drinking water purposes.
- Many modern buildings in the UK use greywater for these purposes.
- Local sustainable water schemes are found in many LICs, however many are beset with problems, such as the Agra clean water project in northern India.
- This scheme is designed to provide water to the poorest areas of Agra, where waterborne diseases are a persistent problem due to polluted sources.
- The solution has been partially solved by bottling clean drinking water from 130 km away, which is funded by UK charities and non-governmental organisations (NGOs).
- The scheme has many issues that limit its success, especially a lack of money for the locals to buy the water, little local technical know-how, and few spare parts for the machinery needed for the scheme to operate.

Dams help to increase water supply

Key Point

Conservation of water is an important strategy to save precious resources.

Quick Test

1. What strategies can be used to obtain water in areas where it is in short supply?
2. What methods of sustainable water management can be used in the home?
3. What are the advantages and disadvantages of using groundwater supplies?

Key Words

water transfer scheme
desalination
water conservation
aquifer
greywater

Energy 1

You must be able to:

- Identify areas of the world experiencing energy security and those suffering from energy insecurity or even energy poverty
- Explain how the exploitation of energy resources has many impacts on supply and the environment.

Energy Supply

- Demand for energy resources is rising globally but supply can be insecure, which may lead to conflict.
- Much of the world relies on **fossil fuels** that are non-renewable; reducing our reliance on these will increase our **energy security**.
- Some countries and regions are energy secure, while others suffer from **energy insecurity**.
- Eastern Europe, including Russia, has large reserves of natural gas and coal, with Russia among the top ten countries for oil and uranium production and thus energy secure.
- Much of the rest of Europe is heavily dependent on energy imports as it has declining fossil fuel supply and has energy insecurity.
- The Middle East and North Africa have large oil reserves but unstable governments often affect how much oil they can export.
- This is affected by the fluctuating price of crude oil or political factors such as the UN trade embargo on Iranian oil from 1995 to 2016.
- At present these countries are energy secure but diminishing reserves could change this status in the future.
- North America has large coal resources but its conventional oil resources are largely now exploited. It is now exploiting non-conventional oil and gas reserves in Alaska, the Alberta Oil Sands and through fracking, but its huge **energy consumption** often outweighs supplies and it is still, to some extent, energy insecure.
- Unconventional energy sources, such as fracking, often cost more to exploit than conventional sources due to the remoteness of the resource, security concerns and industrial processes needed to exploit the raw materials (e.g. oil sands).
- Asia (excluding Russia) has large coal and uranium reserves but has a rapidly increasing demand for oil in particular, outweighing available supplies. The region suffers from energy insecurity.
- Sub-Saharan Africa depends on foreign TNCs to exploit resources.
- Many African people use wood as their primary source of energy and much of the population suffers from **energy poverty**.

There are an estimated 892 billion tonnes of proven coal reserves worldwide – enough to last about 100 years at current rates of production

Key Point

Energy supply is determined by many external factors, both physical and human.

- Energy consumption is increasing throughout the world for a number of reasons:
  - Economic development leads to increased consumption.
  - As rising wealth leads to increased living standards, this in turn leads to more electrical items and technology being used in the home and increased rates of car ownership.
  - As the world's population rises, so does the demand for energy.

Energy Usage

- The impacts of energy insecurity lead to energy supply problems for some regions.
- The variation in fossil fuel prices, especially oil and gas, may mean that the cost of exploitation and production in areas that are challenging will lead to shortages in the supply of these fuels.
- The development of technology has enabled humans to exploit energy sources in remote, difficult and environmentally sensitive areas, such as in deep oceans and polar regions.
- Political factors also affect the supply of energy, such as the conflicts in Middle Eastern countries in the late 20th and early 21st centuries (e.g. Iraq and Syria are heavily involved in the control of oilfields).
- In the 1960s, British companies searched for oil and gas in rocks underneath the North Sea, leading to Britain becoming self-sufficient in oil and gas for a while.
- In Alaska, exploitation of oil sources has taken place in environmentally sensitive and remote areas close to the Arctic Ocean.
- Agriculture uses oil products to power farm machinery, for the transport of goods and livestock, and in agricultural chemicals such as fertilisers and pesticides.
- In recent years, agricultural goods like barley, maize and sugar cane have been used to make **biofuels** that are used as a substitute for oil-based fuels.
- Rising prices for oil mean a higher price for biofuels and agricultural chemicals, making food more expensive.
- Oil is also used as a raw material in many industries such as plastics, packaging and textiles, so rises in the price of oil also lead to a higher price for goods.

Quick Test

1. What regions of the world are energy secure and why?
2. Suggest how energy consumption rates vary in different climates.
3. How does the price of oil affect food prices?

Saudi Arabia, USA and Russia are the biggest oil producers

Key Point

Many countries struggle to obtain energy on a reliable basis at a price that can be sustained.

Key Words

fossil fuels
energy security
energy insecurity
energy consumption
energy poverty
biofuels

Energy 2

You must be able to:

- Describe strategies to increase energy supply
- Show how the extraction of a fossil fuel has both advantages and disadvantages.

Strategies to Increase Energy Supply

- Renewable sources of energy are more environmentally friendly than non-renewable sources of energy, but can be more expensive to generate electricity as the set up costs can be greater.
- Fossil fuels, especially coal and gas, are much more efficient at producing energy than renewables.
- All fossil fuels produce carbon dioxide (CO_2), which contributes to the enhanced greenhouse effect that causes climate change; and some produce other pollutants such as sulphur.
- Many countries, such as the UK and Germany, have used gas to replace coal as a source of energy to make electricity, because gas produces only half as much CO_2 as coal.
- Clean coal technology has enabled coal-fired power stations to produce less pollution and CO_2 by removing the pollutants when the coal is burned. However, in the UK they will all close by 2025.
- Nuclear power is seen by many to be the solution to electricity generation as it is a largely carbon-free source of energy.
- Some countries, such as Japan, are reducing their nuclear programmes due to environmental concerns.
- Other countries, such as the UK, are looking to increase their number of nuclear power stations to reduce greenhouse gas emissions.
- Some countries have large numbers of nuclear reactors that produce much of their electrical energy. For example, France has 58 reactors producing 76% of its electricity.

Key Point

Oil is the most important source of energy in the world at present – its impact stretches far beyond transport.

Cooling towers of a nuclear power plant

Renewable Energy Sources

- There are many strategies to increase energy supply not only by using non-renewable sources, but also by using renewable sources.
- Biomass is the use of plants to produce energy; this can include the burning of wood or the conversion of crops into biofuels such as ethanol.
- Many power stations in the UK have been converted to burn biomass as well as coal.

Key Point

Gas and nuclear power are replacing coal in many countries as a source of energy because they produce fewer greenhouse gases.

- Drax power station in South Yorkshire, originally built to burn coal from the nearby Selby coalfield (now closed), now burns coal, wood pellets and processed straw.
- Selby was the UK's most recently opened coalfield in the 1970s, when coal was used to produce electrical energy in nearby power stations as the price of oil was rapidly rising.
- As coal was identified as a major source of CO_2 and workable seams became exhausted, production fell and Selby began to make losses, resulting in its closure in 2004.
- The coal burned at Drax comes from both the UK and abroad (Australia, Colombia, Poland, Russia and South Africa), while the wood mainly comes from the USA.
- Wind power is the most widely used renewable in the UK, with nearly 7000 onshore and offshore wind turbines producing almost 10% of the UK's electricity.
- The Muppandal wind farm in Tamil Nadu in southern India has over 3000 wind turbines and is one of the largest in the world. It produces 1500 MW – about 20% of India's total wind power production.
- As India is a heavy user of coal, installations such as Muppandal reduce the amount of CO_2 and pollution produced, as well as providing jobs in what was a very poor area.
- Hydroelectric power (HEP) is the most established of the renewables as massive dams in many parts of the world, such as the Three Gorges Dam in China, produce vast amounts of electricity. However, micro HEP schemes now provide electricity in remote rural regions around the world, e.g. Bawan Valley on the island of Borneo.
- Tidal and wave power are forms of HEP that use the sea to produce electricity. Most tidal schemes use barrages that trap the incoming tide and release it when electricity demand is high to drive turbines and produce power.
- A tidal power station has been operational since 2007 in Strangford Lough in Northern Ireland.
- **Geothermal energy** is found mainly in areas where there is volcanic activity (such as Iceland) and the steam produced is used to power turbines to create electricity.
- Solar power has massive potential in countries with high annual amounts of sunshine. Even in the UK, with over 650 000 solar power installations, up to 3% of total electricity generation is possible given the right conditions.

Wind turbines and solar panels

A geothermal power plant

Key Point

Renewable sources of energy are expensive to set up and do not provide as much energy as conventional sources.

Key Words

clean coal technology
biomass
geothermal energy

Quick Test

1. Why is gas preferable to coal for producing electricity in many countries?
2. What is the most important renewable source of energy for the UK? Why?

Energy 3

You must be able to:

- Explain why sustainable energy supplies are important for the future survival of the planet
- Describe how energy usage by individuals and organisations can be designed to be sustainable
- Show how technology can be used to reduce energy usage.

How Can Energy Usage be Made More Sustainable?

- A **carbon footprint** is the greenhouse gas emissions created by an organisation, event, product or individual.
- Carbon footprints in HICs tend to be high because of the lifestyles people lead, with a reliance on fossil fuels, heavy use of electrical devices at work and home, and a diet that relies on food that is imported or uses high levels of inputs derived from fossil fuels.
- In LICs, carbon footprints tend to be low because lifestyles are more sustainable as less fossil fuel is used and food tends to be locally produced.
- Energy conservation is important in protecting precious supplies of energy.
- Energy can be conserved through the intelligent design and building of homes and other buildings like offices and factories. Methods include the use of insulation in loft spaces and walls, double glazed windows and larger windows in south-facing walls.
- Using energy-efficient devices, like kettles that only boil one cup of water, or turning off appliances when not in use (rather than using standby buttons) can also help conserve energy in the home.
- The use of LED light bulbs, washing machines that wash clothes at 30°C or lower and thermostats on central heating systems that can be controlled using a mobile phone, are all ways in which homes and workplaces can be made more energy efficient.
- Transport in cities can be made more sustainable by encouraging people not to use cars. Public transport or bicycles are more sustainable and help to reduce carbon emissions.
- Examples of sustainable public transport measures include the Congestion Charge in central London, where cars are charged for entering the city centre to reduce the amount of traffic and cut down on pollution.
- Most of Greater London is covered by the Low Emission Zone, which charges diesel-powered commercial vehicles that do not meet certain emission standards.
- The introduction of bike lanes and bus lanes can help to speed up public transport in towns and cities.
- Park-and-ride schemes in a number of cities (like Oxford and Cambridge) have helped to reduce traffic in the centres.

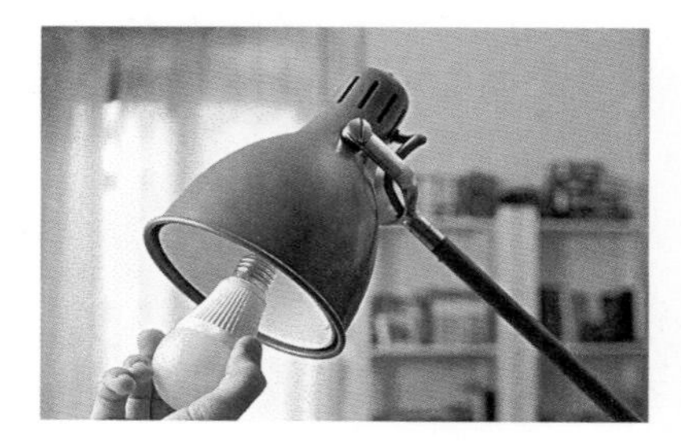

Key Point

Carbon footprints tend to be higher in more developed countries owing to less sustainable lifestyles.

Congestion charging in London

- The re-introduction of trams into city centres like Manchester, Birmingham and Sheffield have helped to ease congestion in these cities and reduce pollution from diesel-powered buses.
- **Hybrid** buses have been introduced in many cities like London, Reading and Brighton, and they combine a conventional engine with an electric one, which results in greater energy efficiency and lower emissions.
- In many UK cities, the provision of bikes to hire at stations and bus stops has been popular following the example of London, which introduced so-called 'Boris Bikes' (named after the then Mayor, Boris Johnson, in 2010). There are now more than 11 000 of these bikes covering the central part of London and other cities like Liverpool have followed suit by introducing bikes for hire.
- Technology has been introduced to increase efficiency in the use of fossil fuels. This includes power stations where clean coal technology has been introduced.
- Cars too have become more energy efficient, with the introduction of new engines that burn fuel more efficiently, hybrid engines in cars like the Toyota Prius and more aerodynamic designs.
- The UK uses about the same amount of oil for transport as it did in the 1970s despite the number of vehicles doubling since then.
- Airliners too have seen similar levels of efficiency with the introduction of aircraft like the Boeing 787 Dreamliner. The Dreamliner uses 25% less fuel than conventional airliners like the Airbus 340 due to efficient engines and the use of lighter materials in the aircraft's construction.
- Education is important in making the public aware of energy conservation.
- Changing people's behaviour (or **adaptation**) in the use of energy is an effective way of reducing usage.
- Examples of energy conservation promotion include the use of posters beside light switches and the teaching of the topic in GCSE courses.

Key Point

It is possible to reduce energy usage through both the actions of individuals and organisations.

Dreamliner aircraft

Quick Test

1. What is a carbon footprint?
2. Suggest ways in which offices may be made more energy efficient.
3. In what ways can urban areas make transport more sustainable?
4. How has technology been used to make transport more energy efficient?

Key Words

carbon footprint
hybrid vehicles
adaptation

Fieldwork

You must be able to:

- Answer questions in an examination showing that you have made links between two fieldwork enquiries about the physical and human geography you have studied.

What is the Geographical Enquiry?

- A **geographical enquiry** is an investigation linking your study in class with fieldwork carried out away from school.
- You will have completed two geographical enquiries in two different places, studying the physical and human geography at these locations.

What Sort of Questions Can I Expect in the Examination?

Questions About How You Might Use Your Fieldwork Techniques in Unfamiliar Contexts

- You may be presented with new information and data about an investigation.
- What techniques were used or would you suggest using for collecting such data?
- What methods were used or would you suggest for presenting the data?
- How effective would these techniques be?

Students measuring orientation of glacial deposits in Snowdonia

Questions About Your Own Enquiries

- You will need to know the titles and be able to describe the location for both of your enquiries.
- Why did you choose the location for your enquiries?
  - Explain why the locations were suitable.
- Why did you choose your question?
  - How is it linked to geographical theories or concepts?
  - Explain any **hypothesis** or hypotheses you tested. A hypothesis is an idea or explanation for something that has not yet been proved. It could be described as an 'educated guess' about the possible outcome of your investigation.
- How did you prepare for your fieldwork?
  - In particular, what risk assessment did you conduct (an analysis of possible dangers) for both enquiries and what measures did you take to reduce these risks?

Field sketching in Carding Mill Valley, Shropshire

- Why did you collect your data?
  - You will need to describe and explain the **primary data** (data you and your fellow students have collected) and the **secondary data** (data collected by another person, group or organisation that you have accessed via articles, websites, books, etc.).
- Is your data **quantitative** (statistics; numbers)?
  - For example, width and depth of a river channel; an environmental quality survey.
- Or is your data **qualitative** (non-numerical; might involve subjective judgements)?
  - For example, field sketches; photographs; video; quotes or opinions.
- How did you collect your data?
- What methods of data collection did you use?
  - Mention **sampling techniques** you used to ensure the reliability of your data, e.g. random; systematic; stratified. If you used **GIS**, explain why such georeferenced data is useful.
  - Justify a method of data collection that you used.
- How did you process and present your data?
- Evaluate a method of data presentation that you used.
  - Describe and explain what you did with your data to make sense of it; what visual, graphical or cartographic methods (including GIS) you used and why.
- What did your data tell you?
  - Describe, analyse and explain your data.
  - What links did you find between data sets?
  - Which statistical techniques did you use?
  - What patterns and anomalies did you find?
- What conclusions did you make?
- Did your conclusions match your expectations at the beginning of the enquiry?
  - Make sure you refer to the original aims or hypotheses. A hypothesis may be correct or incorrect – it is important that you can explain why.
- How effective was your investigation?
  - How could you improve your data collection techniques?
  - What were the **limitations** of your data?
  - How reliable were your conclusions?

Quick Test

1. What is a hypothesis?
2. What is the difference between primary and secondary data?
3. What is the difference between quantitative and qualitative data?

A student measuring river channel gradient in Carding Mill Valley, Shropshire

Key Point

For both of your enquiries, you need to be able to describe and explain what you did at each stage and why. You must also be able to reach plausible conclusions supported by your data, evaluate your work and recognise any limitations.

Key Words

geographical enquiry
hypothesis
primary data
secondary data
quantitative data
qualitative data
sampling techniques
GIS
limitations

Issue Evaluation

You must be able to:

- Apply geographical knowledge, understanding and skills to a particular issue.

What is the Issue Evaluation?

- The issue evaluation encourages **synoptic thinking**, which means applying the knowledge, understanding and skills from all of your geographical learning to a new situation at a range of scales from local, regional, national to international.
- The topic for the issue evaluation will be linked mainly to one of the compulsory (core) topics, but there will also be links to other topics.
- 12 weeks before the exam, a 'pre-release' resource booklet will be published about the issue.
- You can make notes on your resource booklet, but you cannot take this copy into the exam – a new copy will be provided for candidates.
- You will need to show that you understand the issue by referring to the resources.
- You will need to demonstrate **critical thinking** about the issue, which means showing your ability to interpret, analyse and evaluate ideas and arguments.
- **Problem-solving** is required: you will need to make a decision by choosing a possible option and justifying your decision.
- Longer, extended written answers will be expected for some questions.

Key Point

The issue evaluation encourages synoptic thinking. You will have to draw upon knowledge from all of the geography you have studied.

What Geographical Evidence is in the Resource Booklet?

- The pre-release booklet will contain a range of resources, such as different kinds of maps of different scales, GIS data, charts, diagrams, tables of statistics, photographs, aerial or satellite imagery, field sketches and written materials (e.g. web or newspaper articles; quotes from **stakeholders**, **key players** and **interest groups**).
- All the resources are important. You will need to use evidence from them to demonstrate your understanding of the issue and to support the arguments for or against a point of view. It may well be useful to use more than one resource at a time.

Key Point

You will need to take into account a wide range of facts and views to inform your decision-making.

How Can I Reach a Decision?

- Consider these questions:
  - What are the **impacts**? Remember, impacts can be positive as well as negative.
  - What are the main sources of conflict and agreement, e.g. between stakeholders, key players and interest groups.
  - How sustainable is each option or strategy?
  - How can I structure my thinking? Use categories such as advantages versus disadvantages; costs and benefits; SWOT analysis (strengths, weaknesses, opportunities, threats); economic, social and environmental considerations (remember that there are different types of pollution – air, water, noise, light, litter, etc.).
  - Why is **management** important? There are different approaches to governance, e.g. top-down (i.e. decisions are made from the top, such as national government, usually for large-scale schemes) versus bottom-up (i.e. decisions are based on everyone's opinion, such as those of local people, usually for small-scale schemes). Sustainable decisions consult people so they feel empowered (i.e. they feel they have greater ownership of the project and are therefore prepared to invest time and energy in supporting it).
  - How should I reach a decision? Explain reasons for choosing one option and reasons for not choosing other option(s). There will not be a 'right' or 'wrong' answer, but whatever decision you make must be supported by evidence from the resources.

Key Point

There is no 'correct' or 'incorrect' answer, but your decision must be supported by evidence in the resources.

Quick Test

1. What is meant by 'synoptic thinking' and 'critical thinking'?
2. Give definitions, with examples, of:
   a) stakeholders
   b) key players
   c) interest groups.
3. Explain what is meant by 'top-down' and 'bottom-up' approaches to management.
4. What is meant by 'empowerment'?

Key Words

synoptic thinking
critical thinking
problem-solving
stakeholders
key players
interest groups
impact
management

Review Questions

Urbanisation

1. What is urbanisation? [1]
2. Identify two reasons why LIC cities are growing so fast. [2]
3. Which of the following statements are true? Which are false?

   a) In 1900, 40% of the world's population lived in urban areas.
   b) A mega-city has a population of over 10 million people.
   c) Low wages are a 'pull factor' of rural to urban migration.
   d) The majority of the world's population now live in urban areas.
   e) The fastest rates of urban growth are experienced in higher income countries. [5]

4. Decide if each of these refer to **push factors** or **pull factors** of rural to urban migration:

   a) Unemployment
   b) Higher wages
   c) Isolation
   d) Unprofitable farming
   e) Better schools and hospitals [5]

Total Marks .... / 13

Urban Issues and Challenges 1

1. Describe in detail two ways that quality of life can be improved in favelas such as Rocinha. [4]
2. Identify three positive aspects of life in Rio de Janeiro's favelas. [3]
3. Describe one environmental challenge faced by Rio de Janeiro. [2]
4. Describe the living conditions in Rio's favelas. [4]

Total Marks .... / 13

Urban Issues and Challenges 2

1. Describe two environmental challenges faced by London. [4]
2. Describe two transport developments that have taken place within London's Docklands. [4]
3. Give three reasons why tourists are attracted to London. [3]
4. Identify three social challenges faced by London. [3]

Total Marks .... / 14

Urban Issues and Challenges 3

1. Cities can sometimes be described as 'systems'.

   Identify two outputs of a city system. [2]

2. Identify two alternative fuel sources that could provide cleaner energy for transport systems. [2]

3. Explain ways that housing can be designed to be more sustainable. [4]

4. Describe two environmental improvements to urban areas that can form features of sustainable living. [4]

Total Marks .... / 12

Measuring Development and Quality of Life

1. What components make up the Human Development Index (HDI) and why has it become popular as a measure of world development? [3]

2. Explain how a NEE (newly emerging economy) differs from a LIC (lower income country). [2]

3. Identify three population characteristics of LICs. [3]

Total Marks .... / 8

The Development Gap

1. What historic causes have contributed to global inequalities? [2]

2. How can intermediate technology help to reduce the development gap? [4]

3. What is the difference between 'debt reduction' and 'debt relief'? [4]

4. What disadvantages might Kenya experience as it attempts to develop its tourist industry? [4]

Total Marks .... / 14

Review Questions

Changing Economic World Case Study – Vietnam

1. Outline the advantages and disadvantages of TNCs locating in countries with newly emerging economies. [6]
2. Describe how aid can bring benefits to an area. [3]
3. Discuss the impacts of economic development on people and the environment. [5]

Total Marks .... / 14

UK Economic Change

1. Which of these statements describing the UK north–south divide are true? Which are false?
   - **a)** Wages are higher in the north.
   - **b)** The north has traditionally depended on heavy industries.
   - **c)** People have longer life expectancy in the south.
   - **d)** Average house prices are higher in the north.
   - **e)** Unemployment rates are higher in the south. [5]
2. Explain why deindustrialisation has occurred in the UK. [3]
3. Discuss the social and economic changes in a rural landscape that has experienced population growth. [5]

Total Marks .... / 13

UK Economic Development

1. Name the countries that make up the G7. [3]
2. What major transport infrastructure improvements are planned for the UK in future years? [3]
3. Discuss how the UK is linked to the wider world. [5]

Total Marks .... / 11

Practice Questions

Overview of Resources – UK

1. What are 'food miles'? [2]
2. Why is water use increasing in HICs like the UK? [3]
3. What strategies are employed to supply water to regions like south-east England, where demand is greater than supply? [3]
4. Explain why the development of hydraulic fracturing (fracking) potentially has both advantages and disadvantages for the UK. [5]

Total Marks .... / 13

Food 1

1. What is meant by 'famine'? [1]
2. Explain how LICs can improve food security in their countries. [3]
3. List some of the physical and human factors that affect food production. [4]
4. Describe the potential consequences of climate change on food production for the poorest LICs. [4]

Total Marks .... / 12

Food 2

1. What is meant by the term 'irrigation'? [1]
2. What is meant by the 'Blue Revolution'? [2]
3. Explain how any **two** of the following are able to increase food supply:

   hydroponics **aeroponics** **biotechnology** **appropriate technology** [4]
4. For a large-scale irrigation scheme you have studied, describe how its development has both advantages and disadvantages. [5]

Total Marks .... / 12

Practice Questions

Water 1

1. Define the term 'water insecurity'. [2]
2. Suggest two causes of water insecurity. [2]
3. As countries become more economically developed, their use of water increases. Explain why this happens. [4]

Total Marks .... / 8

Water 2

1. What is meant by a 'water transfer scheme'? [1]
2. What is meant by 'desalination'? [1]
3. Describe the main features of a water transfer scheme you have studied. [5]
4. Why is groundwater an important water source for many countries? [5]

Total Marks .... / 12

Energy 1

Study the pie charts below. They show the sources of energy used to make electricity in the UK in 1970 and 2014.

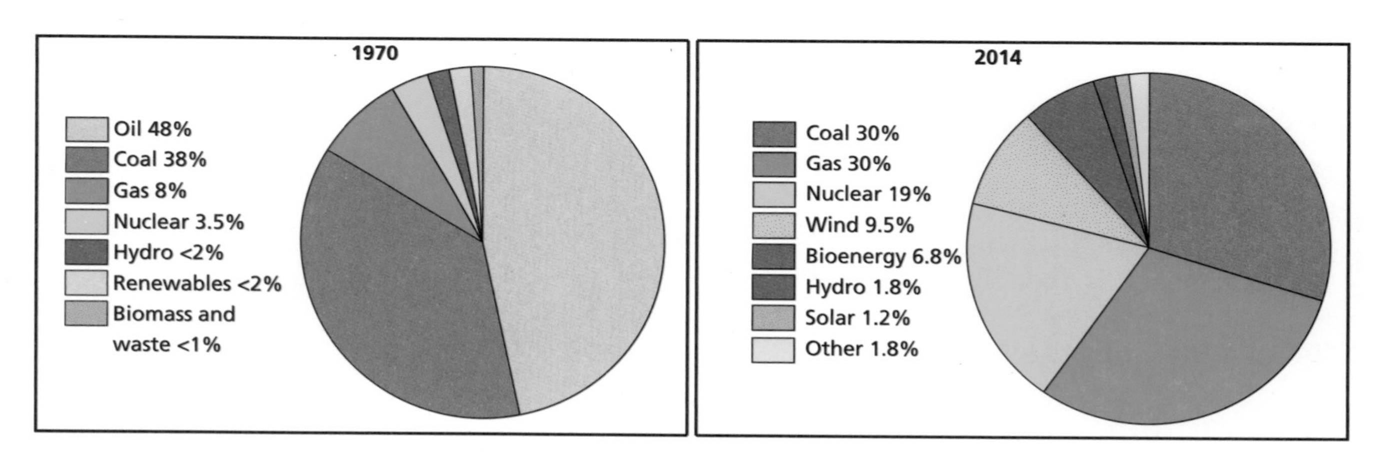

1. Describe the changes in the use of fossil fuels. [3]
2. **a)** What percentage of energy was made up of renewables in 2014? [1]
   b) Why has this percentage increased since 1970? [3]
3. Why are regions like North America and western Europe energy insecure? [4]

Total Marks .... / 11

Energy 2

1. Define 'renewable energy sources'. [2]
2. Why are the owners of coal-powered electricity generating stations having to introduce new technology such as clean coal technology? [3]
3. Describe the main features of a renewable energy scheme that you have studied in a LIC or NEE to provide sustainable supplies of energy. [5]
4. Give some of the advantages and disadvantages of nuclear power. [8]

Total Marks .... / 18

Energy 3

1. Why are carbon footprints high in HICs? [4]
2. Why are carbon footprints low in LICs? [2]
3. How can homes be made more energy efficient? [2]
4. What actions can people take in their homes to cut down on energy usage? [4]

Total Marks .... / 12

Fieldwork

1. Suggest two data collection techniques that could be used to carry out a geographical fieldwork investigation in:

   a) a physical environment [1]
   b) a human environment. [1]
2. Explain some advantages of the locations for each of your fieldwork enquiries. [2]
3. Justify one primary data collection method used in relation to the aim(s) of your physical/human geography enquiry. [3]
4. Explain how a data collection technique could be improved to make the sample more reliable. [3]

Total Marks .... / 10

Review Questions

Overview of Resources – UK

1. What is a carbon footprint? [2]
2. Why does the UK have energy insecurity? [2]
3. Why is the development of hydraulic fracturing (fracking) likely to cause conflicts between different groups of people? [3]
4. How has the way that the UK uses energy changed in the last 50 years? [4]

Total Marks .... / 11

Food 1

1. Define 'food security'. [2]
2. Suggest two different physical causes of food insecurity. [2]
3. Suggest two different human causes of food insecurity. [2]
4. Why do some parts of the African continent have food shortages? [4]

Total Marks .... / 10

Food 2

1. Explain what is meant by 'organic farming'. [2]
2. How can reducing food waste in our homes enable us to be more sustainable? [2]
3. Describe how the 'Green Revolution' was able to increase food production in LICs. [4]
4. Describe the main features of a scheme in a LIC or NEE to increase sustainable supplies of food. [5]

Total Marks .... / 13

Water 1

1. Explain the term 'physical water scarcity'. [1]
2. Explain the term 'economic water scarcity'. [2]
3. Explain how LICs can improve water security in their countries. [2]
4. Why will climate change have an impact on water security for many countries? [3]

Total Marks .... / 8

Water 2

1. What is 'groundwater'? [1]
2. Suggest ways that water supplies can be increased in areas where water is scarce. [3]
3. Explain why the use of groundwater from aquifers is unsustainable. [3]
4. Describe the main features of a scheme that you have studied in a LIC or NEE to increase sustainable supplies of water. [5]

Total Marks ________ / 12

Energy 1

1. Define 'energy security'. [1]
2. Suggest two causes of energy insecurity. [2]
3. Explain why the rising price of oil leads to higher food prices. [5]
4. Suggest reasons why oil resources are being exploited in remote, difficult and environmentally sensitive areas. [5]

Total Marks ________ / 13

Energy 2

1. Define the term 'biofuel'. [1]
2. Why are some countries not suited for electricity production from HEP? [3]
3. What is geothermal energy? [3]
4. What are the advantages and disadvantages of wind power for a HIC such as the UK? [4]

Total Marks ________ / 11

Energy 3

1. What is energy conservation? [2]
2. Describe how the design of buildings can be used to conserve energy. [3]
3. How can transport in cities be made more sustainable? [5]
4. What is clean coal technology? [2]

Total Marks ________ / 12

Review Questions

Fieldwork

1. Describe a data presentation technique you used for your primary or secondary data. [3]

2. Explain why you chose a particular data presentation technique for your fieldwork enquiry. [4]

3. From the photograph, identify the potential hazards and who may be affected by them. [4]

4. Outlining some of your risk assessment, how would you reduce the risks posed by the hazards in your fieldwork? [4]

Total Marks .... / 15

Mixed Questions

1 Study the figure below showing water usage per person in selected countries.

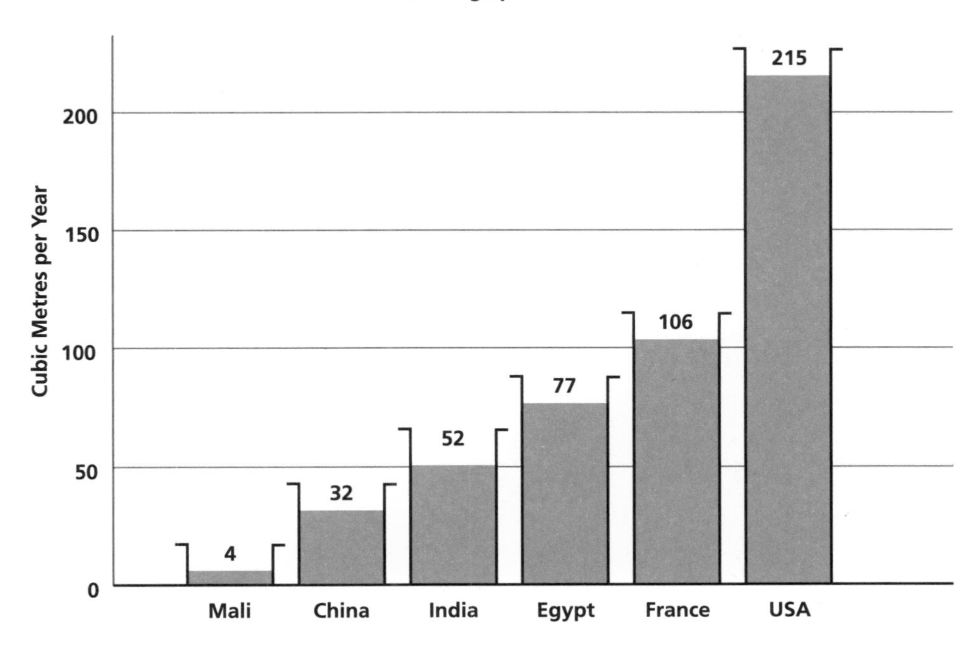

Give reasons for the pattern of water usage shown in the graph. [5]

2 What is the definition of a 'natural hazard'? [2]

3 State two examples of primary impacts of an earthquake. [2]

4 With reference to the Eyjafjallajokull volcanic eruption in 2010, which of the following statements are true and which are false? [5]

a) It occurred at a destructive plate margin. ....

b) It produced a giant ash plume. ....

c) It resulted in serious flooding. ....

d) It caused the cancellation of numerous air flights across Europe. ....

e) It occurred beneath an ice cap. ....

5 Describe how tropical storm intensity is measured on a scale. [2]

....

....

6 How does latitude influence the UK climate? [2]

....

....

7 The chart below is often referred to as the ______________ graph. [1]

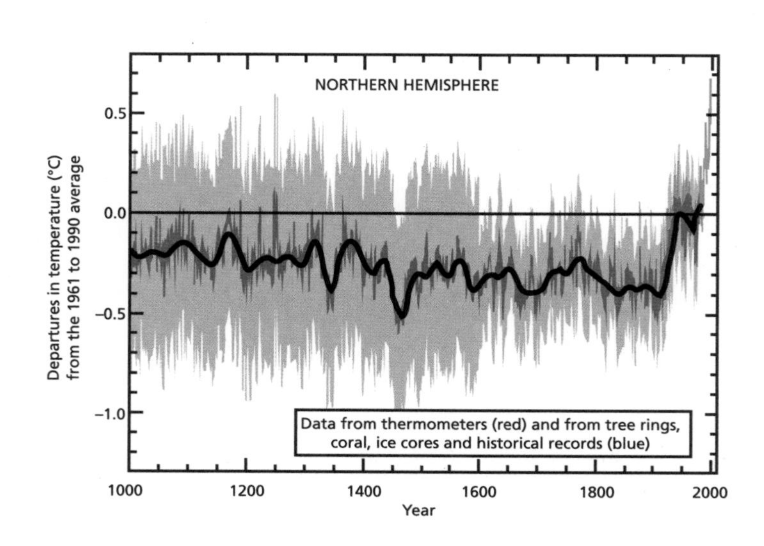

8 Describe one way in which an ecosystem naturally manages itself. [2]

9 Why are beaches said to be temporary features? [3]

10 What word is often used to describe northern coniferous *boreal* forests? [1]

11 What problem can be caused by over-watering soil in desert areas? [1]

12 Give the main factor that led to London's docks closing. [1]

13 Apart from CITES and ecotourism, give two other ways in which rainforests can be sustainably managed. [2]

14 What are the causes and potential impacts of a storm surge with a tropical storm? [4]

15 Why have many terraced fields been abandoned? [1]

Mixed Questions

16. Give an alternative term for freeze-thaw action. [1]

17. How can the landscape influence transport and communication in upland glaciated areas? [2]

18. State two examples of secondary impacts of an earthquake. [2]

19. How can a family use energy more efficiently in the home? [4]

20. Suggest ways in which water resources can be conserved in HICs. [3]

21. Why is the most up-to-date technology not always the best solution to development problems in LICs? [2]

22. Explain the action of abrasion and how it changes a river channel. [3]

23 Diagram A shows the structure of rocks on a cliff.

Diagram B shows what the cliff looks like today.

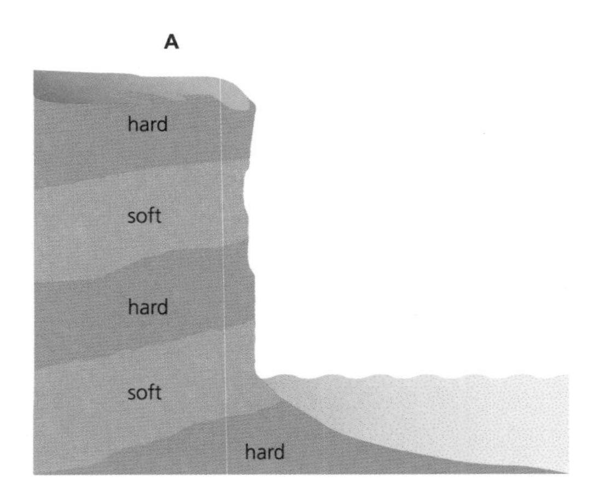

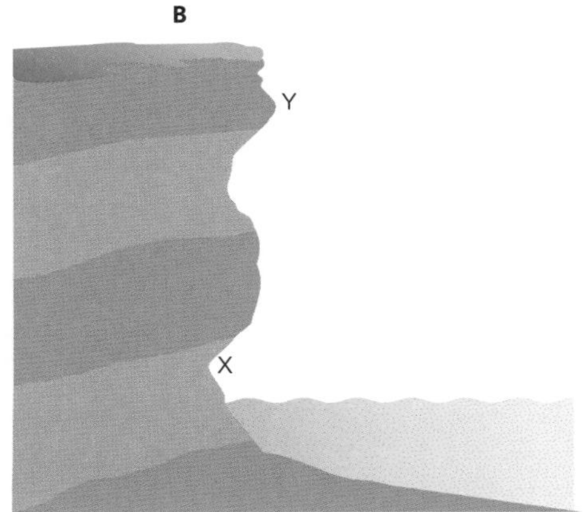

a) What feature is shown at X? .... [1]

b) Describe one physical weathering process that might affect area Y. [2]

....

....

c) What type of mass movement is most likely on this cliff? [1]

....

24 How does calorie intake vary between HICs and LICs? [3]

....

....

....

25 What is the difference between a 'greenfield' and a 'brownfield' development site? [2]

....

....

26 Why can there be waterfalls along the side of a U-shaped valley? [2]

....

....

Mixed Questions

27 On a separate sheet of paper, describe the formation of meanders using only labelled and/or annotated diagrams. [4]

28 In 1910, 10% of the world's population lived in urban areas. What is the UN's predicted figure for 2030? [1]

29 Many local residents opposed the redevelopment plans for London's docks. What might have been the reasons for this? [3]

30 Which three factors are combined to form the Human Development Index (or HDI) to measure development in different countries? [2]

31 What are the solutions that could end the UK's reliance on fossil fuels? [2]

32 How does ethical consumerism make agriculture more sustainable, especially in HICs? [2]

33 Explain the development of an arch from a cave. Refer to named processes. [4]

34 Describe the differences found in carbon footprints in higher income countries and lower income countries. [6]

35 For one of your geography enquiries, to what extent were the results of this enquiry helpful in reaching a reliable conclusion(s)? [9 + 3 SPaG]

Total Marks ______ / 98

Answers

Pages 8–21 Revise Questions

Page 9 Quick Test

1. Destructive boundary – two plates travel towards each other and collide, with the denser plate sinking below the other; Constructive boundary – two plates pull apart to create new land.
2. **Any suitable answer, e.g.** They have always lived there; jobs; confidence in government to 'fix' things; 'it won't happen to me' attitude; fertile soils; valuable minerals; geothermal energy; tourism.

Page 11 Quick Test

1. Primary effects occur immediately while secondary effects occur later on, bringing more problems to those affected.
2. The focus is the point underground where the earthquake originates, while the epicentre is the point on the surface directly above the focus.

Page 13 Quick Test

1. Shield volcano
2. Pyroclastic flow – torrent of hot ash, rocks, gases and steam, moving at up to 450 mph;
   Lahar – 60 mph mudslide of melted snow and volcanic ash.

Page 15 Quick Test

1. Northern
2. Pacific Ocean
3. It can cause coastal flooding, which may lead to more casualties than the high winds.
4. They lose energy supply (warm water) and slow due to friction (especially if the land is hilly).
5. Frequency is how often something happens; intensity is strength and concentration.

Page 17 Quick Test

1. The Philippines
2. Typhoon Haiyan was a Category 5 tropical storm.
3. Primary effects: Landslides occurred across the landscape; storm surges of 5–6 m on the islands of Leyte and Samar; Tacloban Airport terminal was destroyed; the entire first floor of the Tacloban City Convention Center, which was serving as an evacuation shelter, was submerged and many drowned. Secondary effects: economic effects included high losses due to businesses being damaged or closed, and development was halted; social effects included homelessness (1.9 million people), displacement (6 million people), bereavement, disease due to the lack of food, water, shelter and medication, education (schools closed); environmental effects included damage to ecosystems and loss of farmland.
4. Warm surface water in the western Pacific; climate change may have contributed.
5. Immediate responses: much of the central Philippines (Visayas) was placed under a state of national emergency; worldwide relief effort: aid valued at over $500 million.
   Long-term responses: the authorities adopted much more pro-active strategies with 'zero casualty' targets; they tackled the issue of inertia by offering incentives such as free bags of rice to persuade people to leave their homes and property behind.

Page 19 Quick Test

1. The prevailing winds are westerlies from the Atlantic Ocean, bringing relatively warm, moist air from the North Atlantic Drift (or Gulf Stream).
2. a) **Any suitable answer, e.g.** river flooding; sea flooding; winter storms; snow and ice; or drought (e.g. the 2010–12 drought).
   b) **Any suitable answers, e.g. for 2010-12 drought:** Economic and social impacts: farmers struggled to provide water for livestock and harvest crops; low reservoir levels and hosepipe bans affecting six million consumers.
   Environmental impacts: groundwater and river levels very low, affecting aquatic ecosystems; wildfires spread.
   Management strategies: hosepipe bans were introduced; water meters were installed to monitor usage; water companies fixed leaking pipes; water-saving devices were encouraged; education about water use; improved waste-water recycling; new reservoirs and pipelines considered; desalination plants considered.

Page 21 Quick Test

1. Intergovernmental Panel on Climate Change
2. **Any suitable answers, e.g.** ice cores; marine sediment cores; pollen analysis
3. **Any suitable answers, e.g.**
   Physical causes: orbital changes (Milankovitch cycles); volcanic activity (volcanic emissions block sunlight); solar output (changes in the Sun's energy).
   Human causes: use of fossil fuels producing greenhouse gases (e.g. power generation), transportation; agriculture (e.g. methane from cattle, rice paddies); deforestation (e.g. tree removal reduces natural carbon sequestration); methane release from melting permafrost and ocean floors (e.g. due to anthropogenic global warming).
4.

Climate change response	NOT adaptation	NOT mitigation
Electric cars	✓	
Higher sea walls		✓
Tidal power	✓	
Wind farm	✓	
IPCC carbon reduction targets	✓	
Improving air-conditioning in houses		✓

Pages 22–25 Practice Questions

Page 22: Tectonic Hazards 1

1. Convection currents **[1]** in the mantle **[1]**.
2. **Diagram should include at least four of:** convergence, denser oceanic plate, less dense continental plate, melting plate, rising magma, fold mountains. For example:

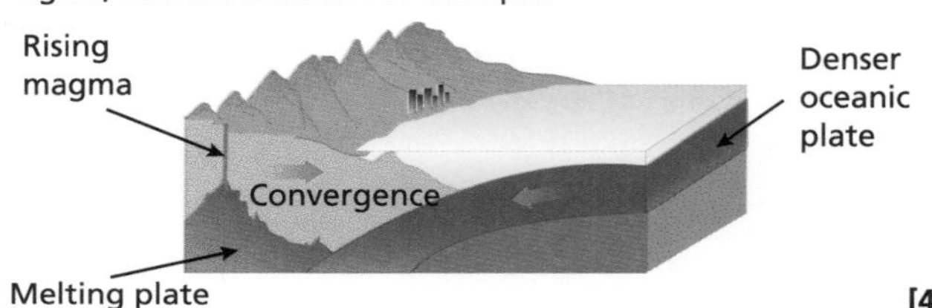

[4]

3. **Diagram should include at least four of:** divergence, rising magma, underwater volcanoes, new rock formed, spreading sea floor, oceanic ridge. For example:

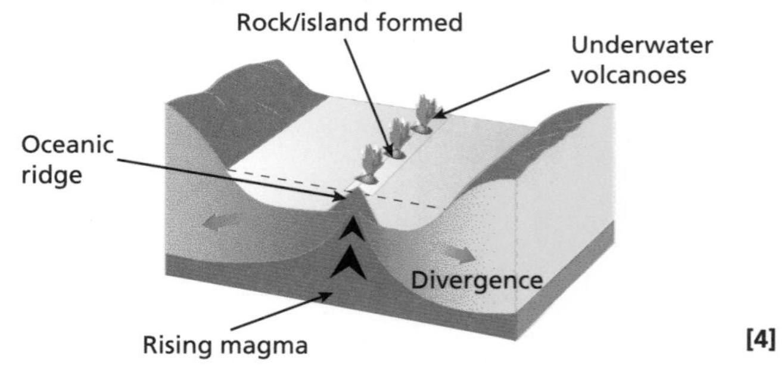

[4]

4. **Diagram should include at least:** direction of movement, friction, faults, crust not destroyed. For example:

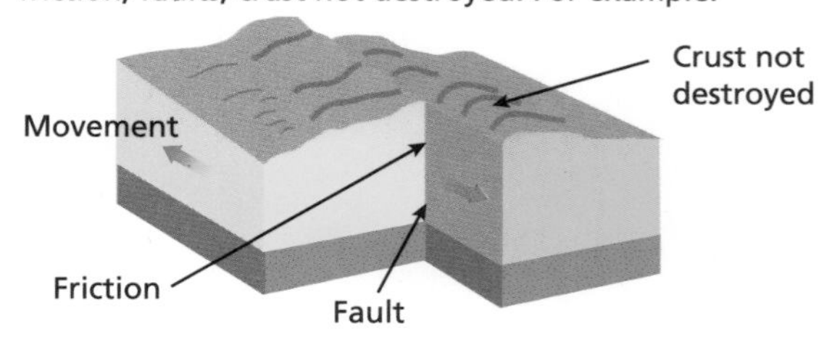

[4]

Page 22: Tectonic Hazards 2

1. LIC – lower income country **[1]**; HIC – higher income country **[1]**
2. A – True **[1]**; B – True **[1]**; C – True **[1]**; D – True **[1]**; E – False **[1]**
3. **Answer may include references to the factors of:** Degree of preparedness, HIC v LIC, population density, quality of buildings,

poverty, depth of focus, etc. **[4]**, and include named examples (such as Japan 2011 – high magnitude and high death toll; Haiti 2010 – lower magnitude and high death toll). **[2]**

Page 22: Tectonic Hazards 3

1. **Any suitable answers with at least two primary effects and two secondary effects:**
   Primary effects included: more than 350 people killed; 360 000 people displaced to emergency shelters; volcanic bombs and hot gases spread for 11 km; pyroclastic flows of up to 9 miles; ash fell up to 30 km away; nearby villages were buried. **[2]**
   Secondary effects included: 350 000 people left homes in the area; sulphur dioxide was blown across Indonesia and as far as Australia; ash cloud disrupted air travel; roads were blocked; lahars; food prices increased; airports closed. **[2]**
2. **Any two from:** Volcanic ash plume at 11 000 metres; fine-grained ash was a hazard to air traffic; lava flows; severe flooding; damage to roads, bridges, water supplies and livestock. **[2]**
3. **Any suitable answer, e.g.** Continued research into improving forecast methods for eruptions; improved warning and evacuation procedures; the construction of safer homes; people moved away from the most vulnerable areas around a volcano; further research into the effects of ash eruptions on air traffic; the construction of dams to hold back lahars; financial support made available to affected farmers and residents. **[Up to 4 marks]**

Page 23: Tropical Storms

1. B **[1]**
2. **Any three suitable features, e.g.** The tropical storm has a circular shape; the cloud is spinning in an anti-clockwise direction inwards towards the centre (which means it is in the Northern Hemisphere); there is a vortex; the storm has an eye. **[3]**
3. a) Cyclone **[1]**
   b) Hurricane **[1]**
   c) Typhoon **[1]**
4. The Coriolis force (Coriolis effect) makes low pressure systems spin anti-clockwise in the Northern Hemisphere **[2]** and clockwise in the Southern Hemisphere. **[2]**

Page 24: Tropical Storms – Case Study

1. Prediction **[1]**; protection **[1]**; planning **[1]**
2. **Any suitable example, e.g.**
   Name: Typhoon Haiyan **[1]**
   Where: The Philippines, south-east Asia (capital: Manila) **[1]**
   When: November 2013 **[1]**
3.

Impacts	Economic	Social/ Political	Environmental
Homelessness		✓	
Factories and other businesses closed or inaccessible due to damage to transport infrastructure	✓		
Waterborne diseases		✓	
Damage to ecosystems			✓
Schools closed for weeks		✓	

[1 for each correct row]

4. a) P **[1]** b) S **[1]** c) S **[1]**
   d) P **[1]** e) P **[1]** f) S **[1]**

Page 24: Extreme Weather in the UK

1. **Any suitable answers, e.g.** 2010–12 drought **[1]**
2. Higher ground in the west has greater precipitation (relief rainfall) and lower temperatures **[1]**. Consequently the east is in a rain shadow, receiving much lower precipitation. **[1]**
3. **Any suitable hydro-meteorological hazards linked to the water cycle, e.g.** River flooding; sea flooding; winter storms; snow and ice; drought **[3]**
4. **Any suitable answers, e.g.** Blocking highs (slow-moving anticyclones) led to very dry winters in 2009–10 and 2011–12 **[1]**. East winds were common, bringing in dry continental air **[1]**. Significantly lower precipitation than normal **[1]**. Climate change may have contributed. **[1]**

Page 25: Climate Change

1. D **[1]**
2. Anthropogenic factors **[1]**
3. **Any suitable causes, e.g.** Use of fossil fuels producing greenhouse gases (e.g. power generation; transportation); agriculture (e.g. methane from cattle; rice paddies); deforestation (e.g. tree removal reduces natural carbon sequestration); methane release from landfill sites. **[3]**
4. Interglacial – Warmer periods of glacier retreat
   Glacial – Extremely cold periods of glacier growth
   Proxy measure – Indirect ways to find out average temperatures from the past
   Quaternary – Last 2.6 million years, including the 'Ice Age'
   [3 if all correct; 2 if two correct; 1 if one correct]

Pages 26–37 Revise Questions

Page 27 Quick Test

1. **Any suitable answers, e.g.**
   Large-scale: tropical rainforest, hot desert, temperate deciduous forest
   Small-scale: pond, hedgerow
2. Food chains show simple relationships between elements within an ecosystem, whereas food webs show complex interrelationships.
3. There will be more food for the prey of that animal.
4. Bird eats seed; bird dies; fungi break down remains; trees reabsorb nutrients from fungi.

Page 29 Quick Test

1. Along the Equator and between the Tropics of Cancer and Capricorn.
2. Mountains, coasts and rivers.
3. Western Europe, north-eastern USA, eastern China, Japan.

Page 31 Quick Test

1. Hot and wet all year round.
2. Trees (e.g. mahogany and kapok) have buttress roots; pitcher plants catch insects.
3. Hot and dry all year round.
4. Saguaro cactus has no leaves to cut down transpiration; quiver tree has fleshy leaves to store moisture.
5. **Any suitable answers, e.g.** Sahara; Kalahari; Mojave.

Page 33 Quick Test

1. **Any suitable answers, e.g.** Mining; settlement; roads; HEP.
2. **Any suitable answers, e.g.** Loss of species; loss of habitats; flooding of forest; soil erosion; locals lose land.
3. **Any suitable answer, e.g.** Selective cutting / logging; ecotourism; labelling schemes; international agreements.

Page 35 Quick Test

1. **Any suitable answers, e.g.** Climate change; population growth; overgrazing.
2. Salts are deposited in the soil surface when water in soils evaporates in high temperatures.
3. **Any suitable answer, e.g.** Encouraging less water use among local people; developing planning laws that restrict the size of buildings; planting trees to stabilise sand dunes; encouraging the use of drip irrigation.

Page 37 Quick Test

1. Cold and dry in winter, mild and dry in summer.
2. Soil with a permanently frozen layer.
3. Benefits – jobs, profits; drawbacks – environmental destruction.
4. Tourists being attracted to new and undiscovered places.

Pages 38–40 Review Questions

Page 38: Tectonic Hazards 1

1. An area that develops above a mantle plume of rising heat from deep in the Earth **[1]**. Magma generated rises and works its way through a thin section of crust to the surface **[1]**.
2. **Any suitable answers, e.g.** Poor quality housing; poor infrastructure making it harder to reach affected people; less money to protect people (e.g. earthquake-proof buildings); less money for responses (e.g. providing food and water); poor healthcare and facilities. **[3]**
3. A – Crust **[1]**; B – Mantle **[1]**; C – Inner core **[1]**; D – Outer core **[1]**
4. **Any suitable answers, e.g.** Building design; monitoring; firebreaks; training of emergency services; education of people in evacuation procedures; planned evacuations; survival kits; overseas aid **[6]**

Page 38: Tectonic Hazards 2

1. Richter scale **[1]**
2. **Any suitable answers with references to the differences between the following in HICs and LICs:** Number of deaths; number of homeless; building damage; effect of communications problems

on people's lives; quality of construction of housing; quality of infrastructure influences ease of access to affected areas; ability to provide healthcare; population density; need for overseas aid. [6]

Page 38: Tectonic Hazards 3

1. **Diagram should include at least four of:** Crater, gentle slopes, low wide cone, lava layers, runny lava. For example: [4]

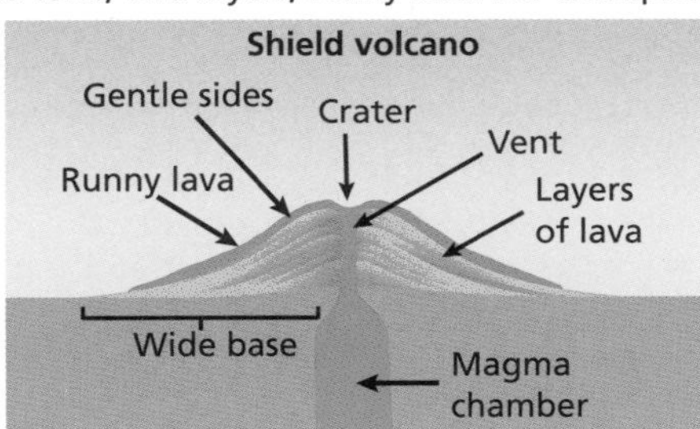

2. Ash cloud – Blocks out the Sun, causing suffocation and health problems
   Lahar – Mudslide including rock debris and water
   Pyroclastic flow – Torrent of hot ash, rock, and gases and steam
   Extinct volcano – Volcano nobody expects to erupt ever again
   Dormant volcano – Volcano that has erupted in past 2000 years but is not currently active
   [4 if all correct; 3 if three correct; 2 if two correct; 1 if one correct]

Page 38: Tropical Storms

1. In excess of 74 mph (119 km / h) **[1]**
2. The intense low pressure **[1]** of a tropical storm creates a 'dome' of seawater **[1]** and a storm surge occurs that causes coastal flooding. **[1]**
3. Warmer oceans expand, so storm surges may be worse **[1]**. Climate change may alter the distribution of tropical storms, and their frequency and intensity may increase **[1]**, but the evidence for this is inconclusive **[1]**.
4.

Country	Name used for tropical storms
Mexico	Hurricanes
Philippines	**Typhoons**
Australia	Cyclones
Bangladesh	**Cyclones**

[1 for each correct column]

Page 39: Tropical Storms – Case Study

1. B **[1]**
2. When people choose to stay in their homes **[1]**, even though they have been warned that they may be under threat from a natural hazard **[1]**.
3. Human factors can make matters worse (3 Ps). Prediction – high level warnings may be delayed (poor quality of governance and communications) **[1]**; Protection – communications infrastructure may be too vulnerable (e.g. it failed in the Visayas in the Philippines) and disease due to the lack of food, water, shelter and medication **[1]**; planning – even though many people may be warned, they may choose to stay in their homes (inertia). **[1]**
4. Foreign investment in new infrastructure, e.g. more resilient bridges and railway – Long-term international response
   Government declares state of emergency across the whole country – Immediate national response
   Worldwide relief effort: aid valued at over $500 million – Immediate international response
   Inertia tackled by offering incentives such as free bags of rice to encourage people to leave their homes – Long-term national response.
   [3 if all correct; 2 if two correct; 1 if one correct]

Page 40: Extreme Weather in the UK

1. **Any suitable answers, e.g.** Economic and social impacts: farmers struggled to provide water for livestock and harvest crops; low reservoir levels and hosepipe bans affected six million consumers. **[1]**
   Environmental impacts: groundwater and river levels were very low, affecting aquatic ecosystems; wildfires. **[1]**
2. **Any three from:** Hosepipe bans; water meters installed; water companies fix leaking pipes; water saving devices; education about water use; improved waste-water recycling; new reservoirs and pipelines considered; desalination plants considered (but high cost). **[3]**
3. The prevailing winds are westerlies from the Atlantic Ocean **[1]**, creating a dominant maritime influence **[1]**, bringing relatively warm, moist air from the North Atlantic Drift or Gulf Stream ocean current **[1]**
4. Place A – Mild winters, cool summers
   Place B – Mild winters, warm summers
   Place C – Cold winters, cool summers
   Place D – Cold winters, warm summers
   [3 if all correct; 2 if two correct; 1 if one correct]

Page 40: Climate Change

1. B **[1]**
2. Milankovitch cycles **[1]**
3. Volcanic eruptions can put enough gas and dust into the Earth's upper atmosphere **[1]** to block out solar energy, which can lead to a volcanic winter. **[1]**
4. Developing more drought-resistant crops or irrigation schemes – A **[1]**
   Making buildings more energy efficient – M **[1]**
   Higher flood defences along coasts and rivers – A **[1]**
   Greater use of renewable resources – M **[1]**
   Carbon capture and storage – M **[1]**
   Planting more trees – M **[1]**

Pages 41–43 Practice Questions

Page 41: Ecosystems and Balance

1. **Any suitable answer, e.g.** Pond; hedgerow. **[1]**
2. A community of biotic (living) **[1]** and abiotic (non-living) components that create an environment. **[1]**
3. Food chains show simple relationships between organisms **[1]**; food webs show more complex relationships. **[1]**
4. Plants grow and are eaten by herbivores **[1]**; the herbivores are then eaten by carnivores **[1]**; the carnivores die and decompose **[1]**; the decomposed remains are reabsorbed by the trees **[1]**.

Page 41: Ecosystems and Global Atmospheric Circulation

1. High pressure **[1]**
2. The movement of air **[1]** owing to many factors, such as the movement of air masses. **[1]**
3. **Any two from:** mountains; coasts; rivers **[2]**
4. They are found in a wide belt encircling the Earth, roughly following the Equator and for the most part between the Tropics of Cancer and Capricorn **[2]**. The Amazon in the northern part of South America and the Congo basin in the centre-west of Africa are the largest rainforests. **[2]**

Page 42: Rainforests and Hot Deserts – Characteristics and Adaptations

1. Latosol **[1]**
2. They provide stability **[1]** and absorb nutrients directly from the fast-decaying leaf litter **[1]**.
3. The climate in tropical rainforests has average daily temperatures of around 27–29°C **[1]**. Rainfall is high at around 500–600 mm each month **[1]**.
4. Fennec foxes are nocturnal, which allows them to avoid the extreme temperatures of the day **[1]**. The fennec also has large ears **[1]**, which act as radiators to lose heat **[1]**.

Page 42: Opportunities, Threats and Management Strategies in the Amazon

1. **Any suitable answers, e.g.** iron; bauxite; copper; nickel; oil **[2]**
2. Only trees above a certain height are felled **[1]**, thus leaving smaller trees to attain maturity **[1]**.
3. **Any suitable answers, e.g.** Hunter-gatherers collect edible plants and catch wild animals to eat **[1]**; soils can be fertilised through small-scale shifting cultivation of crops such as manioc and cassava **[1]**; trees provide fuelwood and building materials **[1]**; medicines and hunting poisons can be extracted from a wide variety of plants and animals **[1]**.
4. **Any two from:** Possible jobs for local people; money brought into the area; energy helps develop local businesses **[2]**

Page 43: Opportunities, Threats and Management Strategies in Hot Deserts

1. Infrastructure **[1]**
2. Hydroelectric power **[1]**
3. Sheep farming has led to the removal of vegetation cover **[1]** through overgrazing and trampling of the soil **[1]**.
4. **Any suitable answers, e.g.** Encouraging less water use among local people **[1]**; developing planning laws that restrict the size of buildings **[1]**; planting trees to stabilise sand dunes **[1]**; encouraging the use of drip irrigation in farming **[1]**.

Page 43: Polar and Tundra Environments

1. High latitudes (60–70° north). [1]
2. Extraction creates huge numbers of jobs [1] but destroys habitats [1].
3. Moderate summer temperatures of 20–25°C [1]. Winter temperatures can plunge below –40°C [1].
4. Plants such as the Arctic poppy [1] survive by developing adaptations such as shallow roots [1] and flowers that track the path of the Sun [1].

Pages 44–61 Revise Questions

Page 45 Quick Test

1. It happens as a result of exposure; no moving agent is involved.
2. Plucking
3. Lowland
4. Jagged / sharp edges

Page 47 Quick Test

1. Truncated spurs; hanging valley / waterfall.
2. Between two corries or along the back wall of corries.
3. Rugged / rocky / lumpy / irregular
4. Mixed sizes and types of rock fragments held in clay.

Page 49 Quick Test

1. Use of farmland or agricultural buildings for non-farming activities.
2. **Any suitable answer, e.g.** Climbing; fell-walking; kayaking; sailing; mountain biking; skiing.
3. **Any suitable answer, e.g.** Water supply to other areas; power supply for iron, textiles, paper production (water wheels or HEP); raw material (brewing); washing (textiles).
4. Sheep farming

Page 51 Quick Test

1. Nearer the sea / away from the cliff
2. Rounded / low / slumped
3. The distance over which a wave has travelled
4. Plunging

Page 53 Quick Test

1. Steep
2. Distal end
3. Stump (or pediment)
4. The area between headlands / the eroded area between headlands.

Page 55 Quick Test

1. Longshore drift
2. To allow water to drain through gently / slowly
3. Hard
4. Reflected away (not absorbed)

Page 57 Quick Test

1. The load
2. It makes them smaller and rounder
3. It makes the channel wider
4. Anywhere along the course of the river

Page 59 Quick Test

1. Gorge
2. Slip-off slope
3. Moving water
4. Fine material deposited by rivers

Page 61 Quick Test

1. Cross-sectional area and velocity of water
2. The difference in minutes or hours between peak / highest rainfall and peak / highest discharge.
3. **Any suitable answer, e.g.** Tarmac; concrete; frozen ground; baked hard ground; some rocks, e.g. slate.
4. Water moving below the surface (e.g. in soil) towards the channel.

Pages 62–63 Review Questions

Page 62: Ecosystems and Balance

1. **Any suitable answer, e.g.** Tropical rainforest; hot desert; temperate deciduous rainforest; boreal / taiga forest.
2. An organism that creates its own food [1] through photosynthesis [1].
3. An organism that eats other organisms [1] and can be herbivore or carnivore [1].
4. An organism that breaks down [1] the dead remains of other organisms [1].

Page 62: Ecosystems and Global Atmospheric Circulation

1. The Equator [1]
2. The Tropic of Cancer [1]
3. The Tropic of Capricorn [1]
4. They are found in a wide belt encircling the Earth in what are known as the high latitudes – areas on and a little south of the Arctic Circle [2]. Much of northern Russia, Canada and Iceland have tundra-type environments [2].

Page 62: Rainforests and Hot Deserts – Characteristics and Adaptations

1. Leaf litter [1]
2. Irrigation [1]
3. They attract insects using scent glands [1] and then catch them using a slippery flower, before digesting them [1].
4. Summer temperatures reach about 40°C and winter temperatures up to about 20°C [1]. Rainfall is low at 3 mm each month [1].

Page 63: Opportunities, Threats and Management Strategies in the Amazon

1. Convention on International Trade in Endangered Species [1]
2. Erosion [1]
3. Commercial means farming for profit [1], whereas subsistence means farming to feed oneself [1].
4. **Any suitable answer, e.g.** Schemes such as that at Tucuruí in Brazil [1] created huge amounts of energy for a growing economy [1] but also flooded huge areas of forest [1]. The dam also encouraged further deforestation for farming and settlement. [1]

Page 63: Opportunities, Threats and Management Strategies in Hot Deserts

1. Hoover Dam [1]
2. Gypsum [1]
3. Small areas of flat land created in hillsides [1], often in the path of natural watercourses [1].
4. Encouraging the use of drip irrigation [1]; implementing appropriate technologies that are cheap and easy to apply by local farmers [1]; recycling water within tourist areas [1]; controlling the size and number of golf courses [1].

Page 63: Polar and Tundra Environments

1. Permafrost [1]
2. Antarctic Treaty [1]; Madrid Protocol [1]
3. The Arctic fox [1] develops a thick coat to protect against the cold [1]. The Arctic hare has small ears to reduce heat loss [1] and white fur to avoid predators [1].
4. **Any three from:** Oil; bauxite; coal; fish [3]

Pages 64–67 Practice Questions

Page 64: Glaciation 1: Processes

1. Terminal moraine [1]
2. At the base and sides [1] of a glacier where it meets the bedrock [1]
3. **Any suitable answer, e.g.**

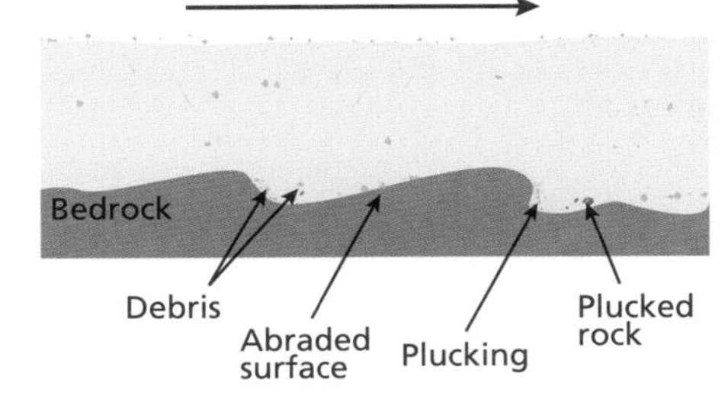

[1 for each correct label]

Page 64: Glaciation 2: Landscape

1. At the beginning / inside edge of a trough / valley [1]
2. **Any two from:** Steep; rocky; frost-shattered [2]
3. Valley floors were made rugged by ice erosion [1]; plucking left hollows [1]; meltwaters washed debris into hollows [1] and rivers are still bringing material down [1] so that a flat valley floor is created and maintained.

Page 64: Glaciation 3: Land Use and Issues

1. **Any two ideas with appropriate explanation from:** High rainfall [1] because of altitude / relief rainfall [1]; steep slopes [1] mean water flows quickly towards valley floors [1]; impermeable rock [1] means that rainwater does not get absorbed / infiltrate [1]; flat valley floors [1] mean that the water does not flow away [1].
2. **Any two features with connected activity from:** Steep, rocky truncated spurs [1] are good for climbing / bouldering [1]; ribbon lakes or corrie lakes [1] can be used for sailing or kayaking / canoeing [1]; high land gets more snow than elsewhere [1] so skiing can take place; mountain scenery attracts walkers [1]; arêtes make routes up to the high peaks for walkers [1].

Page 65: Coasts 1: Processes

1. Loose material is moved up and down the beach by waves [1]. Attrition / the particles crash together [1] so that over time,

edges get rounded and size is reduced.

2. **Any suitable answer which gives a sense of the nature of the rock and cliff process, e.g.** More resistant rock gives steeper cliffs than softer rock. Resistant rock can support higher and steeper land. Softer rock is more easily weathered so the surface is lowered and slumping takes place, giving rounded cliffs. **[4]**
   Or More resistant rock is not easily weathered but particles break away and fall. The cliff moves back / retreats parallel to the original shape. **[4]**

Page 65: Coasts 2: Landforms

1. **a)** Cave **[1]**
   b) Arch **[1]**
2. Flow of water from a nearby estuary **[1]** could prevent further growth so that material is deposited. Second most frequent / dominant waves push the end of the spit round **[1]**.
3. **Any three from:** Strong onshore winds / winds blowing most often from the sea; wide area of sand between high and low tide marks / sand that dries out between tides; mainly sandy beach rather than pebbles or shingle; growth of plants to stabilise the sand (lyme grass, sea couch or marram grass); gently sloping or flat area on landward side of the beach. **[3]**

Page 65: Coasts 3: Management

1. **Any suitable answer with a negative point stated and explained (no credit for description of walls unless used to amplify a point), e.g.** Sea walls tend to reflect wave energy rather than absorb it so there is still energy left to erode / scrape away material at the base of the wall or further down the beach. This makes the base of the wall unstable so likely to fail / need expensive repairs. **[3]**
   Or Sea walls are not attractive / are made of concrete / not made of material like the coast (or similar). The coast attracts people for recreation, especially on the beach but some walls make access to the beach difficult. To make them attractive, with a promenade and gardens on top, costs a lot of money. **[3]**
2. **Any suitable answer making specific comparisons, with emphasis on the techniques employed, their strengths and their weaknesses, e.g.** Nourishment is adding material to a beach that has suffered erosion. This material is usually imported with risks of poor match of size and composition, which can create new problems. It is quite expensive. It usually improves the attractiveness of a beach to people **[2]**. Reprofiling uses local sand / material, redistributing it up the beach to change the shape of the beach so that it should be more resilient. It is usually cheaper. It may not make a beach more attractive to people. **[2]**

Page 66: Rivers 1: Processes

1. Carbonates, e.g. limestone **[1]**
2. **Any one from:** Saltation; traction **[1]**
   Any two points from: Saltation is particles / pieces of rock 'jumping' from the channel bed up **[1]** into the flow of the water then falling back down **[1]**. The impact of this can cause another particle to move **[1]**.
   Or any two points from: Traction is the dragging / rolling **[1]** or sliding **[1]** of material along the channel bed / floor or over other particles **[1]**.
3. **Any suitable answer which covers both ideas and shows understanding that not all eroded particles will be carried, e.g.** Entrainment is the taking in of material from the channel banks or bed **[1]**, which can then be used for abrasion **[1]** to further erode the channel **[1]**. Some eroded material may not be entrained **[1]**, and just falls to the channel bed. **[Up to 3 marks]**

Page 66: Rivers 2: Landforms

1. **a)** A – waterfall **[1]**; B – floodplain **[1]**
   b) Ox-bow lake **[1]**
   c) Source **[1]**
   d) Headward erosion / lengthening of the stream channel **[1]**
2. **Any suitable diagram, e.g.**

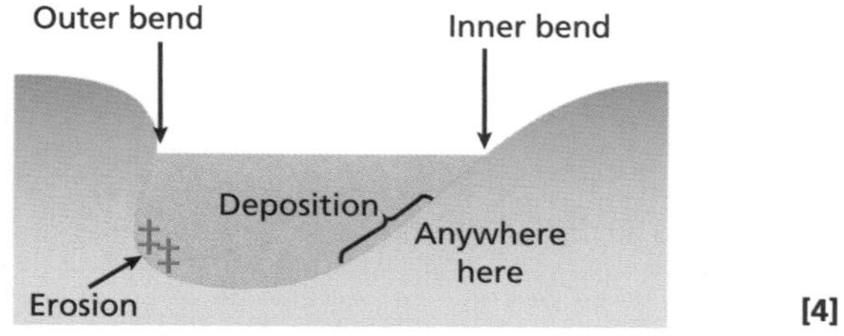

[4]

Page 67: Rivers 3: Flooding and Management

1. **Any two from:** A small catchment area; surrounded by steep slopes; has a lot of tributaries; surrounded by impermeable slopes **[2]**
2. **a)** 2 hours **[1 – unit must be given]**
   b) 50 mm **[1 – unit must be given]**
   c) 4.5 cumecs **[1 – unit must be given]**
   d) 9 **[1]**
3. **Any suitable answer which shows understanding of barriers to water getting into a channel or of those water movement systems that are slower, e.g.**
   Rainwater could be intercepted **[1]** by trees, buildings, etc., then drip to the floor **[1]**, infiltrate / pass into the soil / ground **[1]** and then move towards the channel as throughflow **[1]**.
   Or Rainwater could fall onto permeable ground **[1]**, then infiltrate / pass into the soil **[1]** and continue to travel vertically into the rocks **[1]**, where it will be held and eventually pass laterally / sideways as groundwater flow **[1]** to the channel.

Pages 68–85 Revise Questions

Page 69 Quick Test

1. A city with a population of over 1 million people.
2. 54%
3. Over 10 million

Page 71 Quick Test

1. About 23%
2. **Any suitable answers, e.g.** Favela Barrio Project; upgrading housing; providing pavements; provision of electricity; new sewage systems; legal ownership rights; self-help schemes; low rents; improved transport systems.
3. Local residents are provided with materials like concrete blocks and cement to construct permanent buildings with water and sanitation.

Page 73 Quick Test

1. **Any suitable answers, e.g.** Shops; high-priced apartments; offices; restaurants; yacht marinas; City Airport; water sports; Docklands Light Railway; Thames-side walkways
2. The Crossrail project; HS2
3. The home ground for West Ham United Football Club.

Page 75 Quick Test

1. **Any suitable answers, e.g.** Renewable energy use such as solar panels; insulation; affordable prices and rents; passive energy use.
2. **Any suitable answers, e.g.** Car sharing; park-and-ride; vehicle restriction zones; integrated public transport systems; alternative fuels; cycleways; walkways.
3. A city which produces as much energy as is used (i.e. 'carbon neutral'); a city which supports sustainable lifestyles, such as renewable energy use, local services and jobs, locally-produced food.

Page 77 Quick Test

1. **Any suitable answers, e.g.** Gross national income; life expectancy; adult literacy; birth rate; people per doctor, etc.
2. **Any suitable answers, e.g.** HICs – UK, France, Australia, USA; LICs – Ethiopia, Mozambique, Mali, Democratic Republic of Congo; NEEs – China, India, Brazil, Mexico, South Africa
3. Brazil, Russia, India, China and South Africa

Page 79 Quick Test

1. People move to improve their quality of life (this may be voluntarily or by being forced).
2. Long-term aid is trying to improve people's quality of life, such as access to water, healthcare or education, whereas short-term aid is often in response to a crisis such as providing shelter for people after a natural disaster.
3. Fair trade is where producers receive a guaranteed fair price for the things they make and grow.

Page 81 Quick Test

1. Cheap labour; a growing home market; fewer industrial laws and restrictions; lax environmental laws
2. Multilateral aid is given by countries through an international organisation like the World Bank, whereas bilateral aid is given from one country to another.
3. **Any suitable answers, e.g.** Deforestation; destruction of habitats; air pollution; water pollution; release of greenhouse gases; soil erosion; increase in waste.

Page 83 Quick Test

1. **Any suitable answers, e.g.** Mechanisation of primary industries; depletion of raw materials; competition from overseas; cheaper labour costs overseas; more advanced production methods overseas
2. An area where businesses receive incentives to locate, e.g. tax breaks
3. **Any suitable answers, e.g.** Economic – increased house prices; local shops and services are supported; new businesses set up; investment

into housing improvements; social – schools oversubscribed; traffic congestion leading to longer journey times; services aimed at wealthier residents; community events

Page 85 Quick Test

1. Globalisation is the process by which the world is becoming increasingly interconnected.
2. **Any suitable answers e.g.** Reduced pressure on existing road and train networks; shorter journey times between major cities; creates jobs in the construction industry; encourages economic growth

Pages 86–88 Review Questions

Page 86: Glaciation 1: Processes

1. On top of the ice [1]
2. A – En-glacial moraine **[1]**; B – Terminal moraine **[1]**; C – Sub-glacial moraine **[1]**
3. Deposition means that the glacier does not have enough energy to carry all the load **[1]**. There is less energy if the ice gets thinner **[1]** or if there is melting at the sides and front **[1]**, or if a glacier moves from high to low ground **[1]**.

Page 86: Glaciation 2: Landscape

1. On the floor of a U-shaped valley (trough). [1]
2. Glaciers form in tributary river valleys and flow into the main / spine valley **[1]**. More ice enables more erosion, so more tributaries mean that more ice is available **[1]**.
3. Abrasion on the back wall helps to steepen it **[1]**, then to deepen the floor **[1]**. As ice leaves the corrie, abrasion smooths and shapes the threshold **[1]**.

Page 86: Glaciation 3: Land Use and Issues

1. **Any named upland glaciated area with four features of the climate described, e.g.**
   Snowdonia: wet with about 2600 mm rain per year; cool summer with average temperatures 14–15°C; winter average temperature is 3–4°C but nights and some days are below zero; minimum 60 days of frost; strong winds. [4]
   Or Lake District: wet with 2000 mm rain; cool summer with average temperature 15°C; winter average is about 4°C but colder in the high mountains; over 60 days of frost; strong winds. [4]
   Or Scottish Highlands: wet with 2000 mm rain / snow; very cool summer with average 12°C; winter average is about 1°C; 160 days of frost; good chance of snow; strong winds. [4]
2. **Any suitable answer with linked points, e.g.** Sheep can compact the soil **[1]**, especially if there is overgrazing **[1]**, so that rain does not infiltrate **[1]** and water carries sediment downhill **[1]** into rivers, which can cause flooding **[1]**. Fertilisers to improve soils **[1]** for crops or pasture can wash into lakes or rivers **[1]**, changing the pH **[1]** or affecting native plants and animals **[1]**. **[Up to 4 marks]**

Page 87: Coasts 1: Processes

1. Waves that add material to a beach **[1]**. More effective swash than backwash **[1]**. **[Allow 'stronger' swash or the reverse – weaker backwash]**
2. **Any two from:** Surging waves push material up the beach **[1]** so it will become steeper **[1]** and higher **[1]**. A ridge of pebbles / shingle / material might be left at the top / landward end of the beach **[1]**. **[Up to 2 marks]**
3. **Any suitable answer which describes hydraulic action, the importance of wave action and the results, e.g.** Hydraulic action is a process of erosion. It is very effective on rocks with faults or cracks or joints or visible bedding planes. Approaching waves trap air into the spaces. The air exerts great pressure against the back / sides of the space and loosens rock particles which then fall from the cliff. [3]

Page 87: Coasts 2: Landforms

1. **Any suitable answer which addresses the start of the process and gives a sense of the cliff disappearing but the 'floor' being left (no credit for description, only explanation), e.g.**
   It starts with a notch formed by waves at high water / wave attack low on a cliff / at cliff front. This weakens the cliff / causes the cliff eventually to collapse and retreat at that level. The base of the cliff / underlying rocks are then exposed and eroded mainly by abrasion. [3]
2. **Any suitable answers with a set of ideas that show understanding of the apparent contradiction of weaker bays no longer being eroded, e.g.** Bays form from weaker areas between more resistant headlands but become sheltered by the headlands. As headlands stick out into the sea, wave attack is strong. Waves have further to travel into a bay and have lost energy after breaking on headlands / waves going into bays are low energy, which push sand into the bays / create beaches in the bays. Once a beach is established, waves have further to travel so do not erode the bay any more. [4]

Page 87: Coasts 3: Management

1. **Any suitable answer which addresses the beach as physical protection, e.g.** The further a wave has to travel after breaking, the less erosive power it has. Beaches absorb wave energy / do not reflect it. Material will be deposited, creating even more effective protection. **[3] [Up to 1 mark for any reference to the aesthetic value of beaches, their ecosystem value and economic role in tourism, as long as the point is linked to 'protection'.]**
2. **Any suitable answer, e.g.** Dunes are made of sand grains that are not consolidated / not stable **[1]**. Walking across / sliding down dunes dislodges the sand **[1]** and breaks down the shape of the dune **[1]** so that wind can blow through instead of bringing more sand to increase the dune size **[1]**. Vegetation on dunes is not a complete cover unless they are very old **[1]** and walking will easily break the stems / leaves / blades to stop the plant doing its job **[1]**. Dunes are very attractive, as they have sheltered sections out of the wind, but to reach them people must walk over the ridges **[1]**. **[Up to 3 marks]**
3. **Any suitable answer which shows the idea of change from... to... , e.g.** Before building, the farmland had a long, even width of beach in front **[1]** but after building the beach has partially disappeared **[1]**. Sand / material has piled up against the updrift side / western side of the jetty **[1]** but has been eroded / moved away from the jetty / on downdrift side / to the east of the jetty **[1]**, putting the town at risk **[1]**. **[Up to 4 marks]**

Page 88: Rivers 1: Processes

1. Sand is coarser and so does not lock / stick together **[1]** as well as silt or clay, which are very fine **[1]**.
2. **Any suitable answer which gives some idea of the action of abrasion and then a result, e.g.** Moving water will use the material / particles to scour / scrape the river bank below the surface. The base of the bank will be undercut and the top of the bank may become unstable and collapse. **[3] [Maximum 2 marks for only the process or result]**
3. **Any suitable answer with two points explained, e.g.**
   Drier time / drought **[1]** means there is less water in the channel so its ability to transport material is reduced **[1]**. The channel gets wider **[1]** so there is more friction and a reduction in available energy for transport **[1]**. Channel is less steep **[1]** so less momentum and so less energy available for transport **[1]**. **[Up to 4 marks]**

Page 88: Rivers 2: Landforms

1. **Any suitable answer that makes the link between available energy and water movement, e.g.** Small rivers do not have a great deal of energy available (most is used to move the water along) **[1]** so they take an efficient course around obstacles / wind around obstacles **[1]** because they cannot erode the land / erode sideways **[1]**.
2. [1]

Page 88: Rivers 3: Flooding and Management

1. A diagram / chart showing discharge **[1]** in a river following a rainfall event **[1]**.
2. **Any suitable answer which shows an understanding of the water table and the related ability of an area to hold rainfall, e.g.** A high water table means that the ground will not be able to hold much more water **[1]**, so after further rainfall more water is likely to move towards the channel **[1]**. This may cause the channel to overflow / go overbank / flood **[1]**.
3. **Any suitable answer which states the activity / structures and shows how it was / they were part of the cause of a flood, e.g.**
   York: More impermeable surfaces **[1]** from increase in size of built-up area **[1]** means water gets into channels quicker **[1]**. Water from Yorkshire Dales rivers gets into the Ouse more quickly **[1]** because farming has changed / rainwater isn't being held on farmland as much **[1]**. Erosion of uplands by walkers **[1]** has removed peat which used to hold a lot of water **[1]**. **[Up to 3 marks]**

4. **Any suitable answer which states the place (region or town / city) and gives specific information about a particular storm / flood event, e.g.** York: It is low lying at 15 m **[1]** so water from surrounding highlands could gather there **[1]**. It has two rivers running through it, the Foss and Ouse **[1]** so is doubly vulnerable if discharges rise **[1]**. Upstream from York, the Ouse is joined by the Swale, Ure and Nidd **[1]** – three rivers from the Yorkshire Dales **[1]**, which is an area of high rainfall **[1]**. On December 4 and 5, 2015, the city had flooded, so land was saturated **[1]**. Further rain through the month led to flooding on December 26 **[1]**. **[Up to 4 marks]**

Pages 89–91 Practice Questions

Page 89: Urbanisation

1. a) 10% **[1]** b) 54% **[1]** c) 60% **[1]**
2. The uncontrolled expansion of urban areas (towns and cities) **[1]**. It is considered a problem as it produces deserts of suburban housing **[1]**, reduces productive farmland and destroys wildlife habitats **[1]**.
3. a) **Any three from:** Unemployment; low wages; farming is difficult and unprofitable; few job opportunities; lack of social amenities; isolation; natural disasters. **[3]**
   b) **Any three from:** More job opportunities; higher wages; better schools and hospitals; better housing and services (like water, electricity and sewerage); better social life; better transport and communications. **[3]**
4. **Any suitable answer which focuses on four differences, e.g.** The rate of growth (i.e. fastest in LICs); rural to urban migration (reference to push-pull factors); industrialisation (as countries industrialised, people began to migrate to cities for jobs; as this happened in HICs over 100 years ago, they now have large urban populations; LICs are still in the early stages of industrialisation, and so urban populations are still growing); growth of shanty towns (on periphery of LIC cities, but not a feature of HIC cities); primate cities in LICs (LICs often have one primate city rather than a number of large cities of similar size). **[Credit can be given for use of correct geographical terminology, e.g. HIC, LIC, NEE, millionaire cities (growing rapidly in LICs), mega-cities (mainly found in LICs). Credit can be given for relevant examples.] [Up to 4 marks]**

Page 89: Urban Issues and Challenges 1

1. **Any suitable answers, e.g.** Better healthcare; better schools; better housing; jobs; wages; better social services **[4]**
2. **Any suitable answer, e.g.** Lack of clean water supply; overcrowding; lack of job opportunities; crime rates; lack of services **[4]**
3. **Any suitable answer, e.g.** Beaches; Sugar Loaf Mountain; statue of Christ the Redeemer; Rio Carnival; sports events **[4]**
4. **Any suitable answer which refers to two different social challenges [2], with further development of each point needed [2], e.g.** Migration (rapid growth of Rio in recent years from the movement of people into the city from surrounding rural areas putting pressure on services and amenities); housing (many rural migrants begin their life in the city in shanty towns called favelas; these are unplanned and spontaneous, often growing up on poor quality land; favelas are overcrowded, residents have no legal ownership, and houses are built of cheap materials; these are areas lacking clean water and sewage disposal, with no schools or hospitals, and few job opportunities); healthcare (favelas have high levels of disease and illness, with poor quality healthcare – particularly for maternity and care of the elderly); education (nearby schools suffer from poor funding, low enrolment due to poverty, and a lack of trained teaching staff); water supplies (clean water is not available for 12% of the population in Rio's favelas); energy (a shortage of available electricity supply means frequent blackouts for residents); crime (street crime is high, with powerful drugs gangs controlling the favelas).

Page 89: Urban Issues and Challenges 2

1. **Any two from:** By 1980, ships had become too large to sail up the River Thames **[1]** and the docks became derelict **[1]**; competition from European ports such as Rotterdam **[1]**.
2. **Any suitable answers, e.g.** Housing; sports facilities; improved transport links; the Queen Elizabeth Park **[4]**
3. **Any suitable answers, e.g.** Wider use of diesel-electric hybrid buses **[1]** and the trialling of hydrogen fuel cell buses and combined diesel and biofuel buses **[1]**.
4. **Any suitable answers, e.g.** The Docklands Light Railway (DLR) **[1]** connected the area to the London Underground system **[1]**; London City Airport **[1]** specialises in STOL (short take-off and landing) air travel **[1]**; high speed rail link (HS2) from London to Birmingham (and eventually Leeds and Manchester) **[1]**; Crossrail project from Paddington Station to Reading in the west and Abbey Wood in the east of the city **[1]**; plans for a new runway at Heathrow **[1]**. **[Up to 4 marks]**

Page 90: Urban Issues and Challenges 3

1. 90% **[1]**
2. **Any suitable answers, e.g.** Food and water; fuels and energy; building materials and consumer goods. **[2]**
3. **Any suitable answer which refers to three different strategies [3], with further development (but not necessarily explanation) of each point needed [3], e.g.** Car sharing (resulting in fewer vehicles on the road; it is particularly effective when work colleagues share, as rush hour traffic is reduced); vehicle-restricted areas (that reduce the number of vehicles on roads and reduces vehicle movements in busy, congested areas); increased use of public transport (including buses, trams and trains, organised into an integrated transport system; this helps to reduce the number of vehicles on the roads); bus lanes (improve traffic flow); park-and-ride systems (reduce vehicles on roads in busy, congested areas, and improve traffic flow by reducing need for parking); increased bicycle use (reduces the number of vehicles on roads); new road designs and traffic controls (such as one-way systems and traffic controls to improve traffic flow). **[Up to 6 marks]**
4. **Any suitable answer which refers to a range of different strategies [up to six for 6 marks], with further development of individual points needed to gain additional marks (up to a maximum of 6 in total). A suitable answer could include a variety of points that help a city to become carbon neutral (produces as much energy as it consumes):** Renewable energy for housing (e.g. solar panels – to generate electricity and reduce carbon emissions); reduce energy for industry (e.g. solar panels, wind generators, hydro-electricity – to generate electricity and reduce carbon emissions); reduce energy use at home and in the workplace (through the use of renewable energy sources, efficient insulation, double and triple-glazing, water and electricity meters, low-flow taps, etc.); reduce household and industrial waste (through increased recycling of waste materials, e.g. cardboard, metals, glass, food waste); energy from waste (composting of food and green waste helps reduce methane emissions); use of greywater systems.

Page 90: Measuring Development and Quality of Life

1. **Any suitable answers, e.g.** Birth rate; life expectancy; literacy; infant mortality; access to healthcare **[2]**
2. **Any suitable answers, e.g.** Gross domestic product (GDP); wages; trade figures **[2]**
3. **Any four from:** Stable growth rate; low birth rates; low death rates; ageing population; some HICs may even have declining populations **[4]**

Page 90: The Development Gap

1. **Any two from:** Extreme climates make food production difficult in some areas; land in some regions is unsuitable for farming (e.g. too high, too steep, thin soils); limited water supplies in some regions; natural hazards (e.g. tropical storms and earthquakes) affect some regions more than others; an imbalance in the location of natural resources (e.g. precious metals, energy sources). **[4]**
2. **Any suitable answers, e.g.** A fair price for goods produced (e.g. cocoa) means more money to spend on food, schooling and improving family life **[2]**; co-operatives provide extra services (e.g. healthcare) **[2]**
3. **Any suitable answers, e.g.** Wildlife safaris (the 'big five'); tribal culture; warm climate with all-year sunshine; varied landscapes; spectacular national parks **[4]**
4. **Any suitable answers, e.g.** Tourism provides 15% GDP; creation of tourism-related jobs; efforts made to preserve tribal cultures; better protection of landscapes; greater development of infrastructure **[4]**

Page 91: Changing Economic World Case Study – Vietnam

1. **Any suitable answers, e.g.** Cheap labour; longer working hours; fewer health and safety laws; less stringent environmental laws **[3]**
2. Profits from a transnational corporation are sent back to the home country. **[1]**
3. **Any suitable answer, e.g. for Vietnam:** Approximately 50% of people work in the primary sector **[1]**, in jobs such as farming, mining and forestry **[1]**; around 20% of people work in the

secondary sector in manufacturing [1], such as for companies like Nike making clothes and footwear [1]; around 30% of people work in the tertiary sector in services [1], such as in education, health, shopping [1]; tourism is a rapidly growing tertiary industry in Vietnam [1]. **[Up to 3 marks]**

Page 91: UK Economic Change

1. **Any suitable answers, e.g.** Bristol [1]; Cambridge [1]
2.

2001	2015 – higher, same or lower?
Primary = 1%	Same [1]
Secondary = 24%	Lower [1]
Tertiary and Quaternary = 75%	Higher [1]

3. **Any suitable answer, e.g.** Improvements to transport like HS2 [1], which will make the north a more attractive place to live [1]; it will also attract business investment [1], creating jobs [1]. The Government has set up enterprise zones [1] to attract business to locate in the north [1], creating jobs and economic growth [1]. The Government is giving money and decision-making powers to northern cities [1], ensuring that decision-making is suitable for the economy and people in the north [1]. **[Up to 3 marks]**

Page 91: UK Economic Development

1. **Any suitable answers, e.g.** Germany [1]; USA [1]; China [1]; EU [1]
2. **Any suitable answer, e.g.** Using renewable sources of energy like wind and solar [1] to reduce carbon dioxide emissions [1] (which contribute to the enhanced greenhouse effect and climate change [1]). Reducing packaging [1] means there is less waste to be processed [1]. Designing buildings that blend in with the environment [1], e.g. Adnams in Suffolk has a green roof and lime / hemp walls [1].
3. A greenfield site is a plot of land that has not been used before [1], whereas a brownfield site is land that has been previously used for development [1].

Pages 92–111 Revise Questions

Page 93 Quick Test

1. The UK is unable to grow enough food for its population and there is a great demand for food out of season.
2. Because of a growing population and lifestyles that use more water.
3. It is the driest part of the country but also the area with the highest population.
4. Declining fossil fuel reserves (e.g. gas) mean that the UK has to import more, while more affluent lifestyles mean that households use more energy.

Page 95 Quick Test

1. A country having reliable access to a sufficient quantity of affordable, nutritious food to feed its people.
2. **Any suitable answers, e.g.** Aridity; floods; extreme cold / heat; water availability; soil type
3. **Any suitable answers, e.g.** Civil wars; poor techniques leading to degradation and desertification; poor distribution networks; lack of adequate storage.

Page 97 Quick Test

1. **Any suitable answer, e.g.** The use of new farming methods, such as hydroponics and genetically-modified crops, that supply larger yields; in LICs, examples include high-yielding varieties of cereals and modern farming techniques such as irrigation, synthetic fertilisers and pesticides.
2. **Any suitable answer, e.g.** Eating sources of food that are locally grown and only eating fruit and vegetables that are in season; changing diets to those less reliant on meat and dairy products; using sustainable fish sources and reducing food waste in homes and shops.
3. Ethical consumerism is the purchase and use of sustainable goods, such as food that is locally grown and only eating fruit and vegetables that are in season.

Page 99 Quick Test

1. The ability of a population to ensure access to quantities of safe water to maintain life, social well-being and economic development.
2. **Any suitable answer, e.g.** Greater demand from agriculture, industry and domestic properties; climate change; over-extraction by man.
3. Poor sanitation leads to the pollution of water supplies (this problem increases with a rise in population).

Page 101 Quick Test

1. Methods of conservation can be used in the storage of water; expensive schemes, such as water transfer schemes and desalination plants, can be used.
2. **Any suitable answers, e.g.** Using low-flush toilets; taking short showers; only using washing machines and dishwashers when full.
3. **Any suitable advantages, e.g.** The quality of the water is very good and needs less treatment than river water to make it safe to drink; it stays available during the summer and during droughts when rivers and streams have dried up.
   Any suitable disadvantages, e.g. The use is not sustainable as the groundwater takes a long time to replenish.

Page 103 Quick Test

1. Eastern Europe, including Russia – have large reserves of natural gas and coal; Middle East and North Africa – have large oil reserves.
2. Energy usage is high in countries that have either hot or cold climates due to heat or air-conditioning use; usage is also high in countries with a great annual range of temperature.
3. Agriculture uses oil to power machinery, for transport and in agricultural chemicals, so an increase in oil price causes a rise in food prices.

Page 105 Quick Test

1. Gas produces only half as much carbon dioxide as coal and, with the introduction of new exploitation methods like fracking, is relatively cheap.
2. Wind power is the most widely used renewable in the UK because it experiences windy conditions for much of the year. Nearly 7000 onshore and offshore wind turbines produce almost 10% of the UK's electricity output.

Page 107 Quick Test

1. The greenhouse gas emissions caused by an organisation, event, product or individual.
2. **Any suitable answers, e.g.** The use of LED lighting; motion sensors on lighting to turn it off when no-one is around; double glazing and large south-facing windows to let in more light.
3. **Any suitable answers, e.g.** The introduction of park-and-ride schemes; bus and cycle lanes; congestion charging; trams; hybrid buses; bikes for hire.
4. **Any suitable answers, e.g.** The use of hybrid and more efficient engines in vehicles; electric vehicles; aircraft that are more fuel-efficient; vehicles that are more aerodynamic.

Page 109 Quick Test

1. A hypothesis is an idea or explanation for something that has not yet been proved.
2. Primary data is information collected by you and your fellow students. Secondary data is information you have found on a website or in a book (i.e. which someone has collected previously).
3. Quantitative data includes statistics and numbers (i.e. things that can be counted or calculated). Qualitative data is non-numerical and can be subjective (i.e. things such as field sketches, opinions, news articles).

Page 111 Quick Test

1. Synoptic thinking means using the knowledge, understanding and skills that you have learned throughout the course to help answer questions about a new situation that you haven't studied before. Critical thinking is interpreting, analysing and evaluating ideas and arguments based on the information that you are given and your existing knowledge and skills.
2. a) Stakeholders are people or groups with a real interest in an issue or in the development of a decision or policy, e.g. the residents of an area where a new road is to be built.
   b) Key players are people or groups with an interest and a significant degree of control about an issue or in the development of a decision or policy, e.g. the company paid to build the road, or the politician who gives it the go-ahead.
   c) Interest groups are organisations of people with a common cause about which they attempt to influence policies or decisions, without seeking political control, e.g. if the residents of the area where the new road is to be built formed a group, it would be an interest group.
3. 'Top-down' management is where the person / organisation in charge makes all the decisions and tells those lower down exactly what to do. A 'bottom-up' approach takes into account everyone's opinion and the decision is formed by the whole group.
4. Measures that provide people with a degree of control over problem-solving or decisions.

Pages 112–114 Review Questions

Page 112: Urbanisation

1. **Any simple definition of urbanisation, making reference to an increase in the proportion of people living in urban areas (or towns / cities / built-up areas).** **[1]**

2. **Any suitable answer with two different reasons. A suitable answer should refer to any 'pull' factor of rural to urban migration, such as industrialisation creating a demand for jobs, better healthcare, better schools, etc. or any 'push' factor of rural to urban migration, e.g. unemployment, difficulties in farming, low wages, etc.** **[2]**
3. **a)** False **[1]** **b)** True **[1]** **c)** False **[1]** **d)** True **[1]** **e)** False **[1]**
4. **a)** Push factor **[1]** **b)** Pull factor **[1]** **c)** Push factor **[1]** **d)** Push factor **[1]** **e)** Pull factor **[1]**

Page 112: Urban Issues and Challenges 1

1. **A suitable answer could include:** Provision of housing for rural migrants (instead of allowing uncontrolled growth of favela housing); upgrading of favelas (by providing pavements, electricity and sewage systems through schemes like the Favela Bairro Project); promotion of self-help building schemes (where local residents are provided with building materials); giving residents legal rights of ownership or low rents on properties (to develop permanent communities); improved transport systems (to give residents better access to work in the city); encouraging creation of businesses like shops and restaurants (to provide employment and allow favelas to be self-contained); improved law and order through pacification programmes (to reduce crime in the favelas); encouraging tourism and creation of tourist-related businesses (to bring additional income opportunities to residents). **[Two different ways that quality of life can be improved need to be identified [2], with further development required [2]]**
2. **A suitable answer could include:** Strength of community; creation of new businesses (e.g. shops, restaurants, tourist-related businesses, cottage industries like pottery); residents sending money home to families in the countryside; recycling (with particular reference to use of building materials and waste recycling); cultural and ethnic diversity; growth in tourist-related activities. **[Three different positive aspects need to be given [3]]**
3. **A suitable answer could include:** Urban sprawl (as the city continues to grow, it encroaches on surrounding rural areas); pollution (air pollution from heavy traffic); traffic congestion (in the city centre); sea pollution (from sewage and industrial waste); waste disposal (particularly in the favelas, many of which are inaccessible to collection vehicles). **[Identification of one environmental challenge [1], with further development required [1]]**
4. **A suitable answer could include:** Favelas being unplanned and spontaneous (often growing up on poor quality land); overcrowding (with many houses crammed into a small area); poor quality housing (often built from cheap materials like wood or corrugated iron); lack of services (like clean water, sewage disposal, electricity supply); poor health (due to lack of clean water and lack of sewage disposal); few job opportunities (leading to poverty); few facilities (such as schools, hospitals and public transport); high levels of crime (particularly street crime, with powerful drugs gangs controlling many areas). **[Reference needed to a range of different living conditions for up to 4 marks), with further development of individual points needed to gain additional marks (up to a maximum of 4 in total).]**

Page 112: Urban Issues and Challenges 2

1. **A suitable answer could include:** Dereliction (as manufacturing industries have declined, particularly in the inner city, much land has been left in a state of dereliction); waste disposal (a large city produces a lot of household and commercial waste for disposal); pollution (atmospheric pollution from industry and vehicles). **[Two different environmental challenges identified [2], with further development of individual points [2]]**
2. **A suitable answer could include:** The Docklands Light Railway (DLR – which connected the area to the London Underground system); London City Airport (a STOL airport providing quick access to the city business areas); improved public transport (new underground station next to the Olympic site). **[Two different transport developments identified [2], with further development of individual points [2]].**
3. **Any suitable answer, e.g.** Historic buildings; monuments; sporting events; cultural events; entertainment (credit can be given for named relevant examples). **[Up to 3 marks]**
4. **A suitable answer which includes three social challenges, e.g.** Housing inequalities (provision of more affordable housing); need to improve education standards to correct variation in performance between boroughs; poor access to healthcare for people with poor English language skills; increased in-migration (particularly international migration) attracted by finance and knowledge-based employment opportunities. **[Up to 3 marks]**

Page 113: Urban Issues and Challenges 3

1. **Any two from:** Sewage; exhaust gases; household waste; industrial waste; building waste. **[2]**
2. **Any suitable answers, e.g.** The use of hydrogen-fuelled **[1]** and electric cars **[1]** helps to reduce air pollution and cuts down on reliance on fossil fuels.
3. **Any suitable answer with development of each point made, e.g.** Renewable energy use such as solar panels (which reduces the requirements for energy from fossil fuels and reduces carbon emissions) **[1]**; insulation (of walls, lofts, etc. which reduces heat loss and therefore reduces energy use) **[1]**; affordable prices and rents (which supports a balanced community of different income levels, and reduces demand on local authority housing) **[1]**; passive energy use (to reduce demands on energy produced from fossil fuels and reduce carbon emissions) **[1]**.
4. **A suitable answer could include:** Urban greening (a target of 40% green space of parklands and trees help to meet leisure needs of residents and absorb carbon dioxide from the atmosphere); urban forests (e.g. Adelaide in Australia – they help to meet leisure needs of residents and absorb carbon dioxide from the atmosphere); urban architecture (such as vertical farming, allotments and roof-top gardens help to increase food production and reduce 'food miles' from imported products); use of brownfield sites for development (to avoid the need to expand into untouched greenfield sites at the edge of the urban area); greenbelt (to maintain an area of green space in easy reach of all residents within the urban area). **[Two different environmental improvements need to be identified [2] with further development of individual points needed [2]]**

Page 113: Measuring Development and Quality of Life

1. HDI combines life expectancy **[1]**, literacy and income to measure development **[1]**. It has become popular as it does not rely on one single indicator (usually wealth). **[1]**
2. **Any suitable answer with two clear reasons to show how a NEE differs from a LIC, e.g.** Country wealth (a LIC is classified by the World Bank as one with less than $1045 GNI per capita); rate of economic development (NEEs are beginning to experience high rates of economic development, usually with rapid industrialisation); reliance on agriculture (NEEs no longer rely primarily on agriculture in their economy. (Credit can be given for naming relevant examples.) **[Up to 2 marks]**
3. **Any three from:** Population increasing rapidly; high birth rates; falling death rates; many young dependents **[3]**

Page 113: The Development Gap

1. **Any suitable answers, e.g.** Corrupt governments **[1]**; war and conflicts **[1]**
2. **Any suitable answers which include:** A description of intermediate technology (a form of aid, sometimes also known as 'simple' or 'appropriate' technology that does not necessarily use the most up-to-date technology available to improve a situation in a LIC) **[2]**; and for identifying how intermediate technology can reduce the development gap (the difference in standard of living between rich and poor countries) through making improvements to living or working conditions in a LIC **[2]**. (Credit can be given for reference to examples of intermediate technology aid projects.)
3. Debt reduction occurs when countries are released from some of their debts **[1]**, although often with other conditions attached **[1]**. Debt relief is when lending countries write off debts **[1]** that have spiralled out of control due to high interest payments. **[1]**
4. **A suitable answer could include:** Money generated goes to large overseas companies (rather than benefiting local people); tribes forced off their land to create national parks (causing changes to lifestyles and loss of cultural integrity); dress and behaviour of tourists can offend local people (particularly certain local religions); safari vehicles increase in number and drive off-road to locate wildlife (disturbing wildlife and damaging natural vegetation); coral reefs damaged by divers and tourist boats (causing destruction to reefs which then fail to regenerate); threat of terrorism **[4 marks for identifying four different disadvantages, or fewer disadvantages can be given with additional marks added for development of individual points]**

Page 114: Changing Economic World Case Study – Vietnam

1. **A suitable answer could include:**
   Advantages: secondary sector employment opportunities (particularly for women, providing new skills and higher wages); cheap labour supply (for the TNC); growth in the home market (adding to exports and making a positive contribution to GDP); fewer industrial laws and restrictions (makes operations for the TNC easier); extra tax revenue (that can be spent by the host country on infrastructure improvements); lead to attraction of other TNCs ('snowball' effect brings strength to the host country's economy). **[Up to 3 marks]**
   Disadvantages: company image and advertising (can undermine culture of host country); political influence held over host country government (can lead to undemocratic decisions); investment easily moved away from host country (can leave unemployment and a vacuum in the host country economy); environmental impact (pollution from factories). **[Up to 3 marks]**
   [Advantages and disadvantages could apply to either a TNC itself, or the host country. Marks can be awarded for each individual factor identified, and additional marks can be gained for further development of points made.]
2. Improving transport infrastructure enables people to access services like education and healthcare more easily **[1]**. It also encourages businesses to develop **[1]** and can attract foreign transnational investment **[1]**. Training workers provides people with new skills **[1]**, enabling them to get work **[1]** and improve their quality of life **[1]**. Promoting women's economic empowerment encourages women into the workforce **[1]**, enabling them to earn money and improve their quality of life **[1]**. It also raises the status of women **[1]**. Tackling hunger means people are healthier **[1]**. This reduces demand for healthcare **[1]** and people can work and earn money **[1]**. **[Up to 3 marks]**
3. Industrial development can cause air pollution **[1]** and can affect people's health, e.g. asthma **[1]**. Waste and toxic chemicals cause land and water pollution **[1]**, which can affect public drinking water supplies **[1]**. Increasing the use of fossil fuels releases greenhouse gases that cause the enhanced greenhouse effect **[1]**. This causes climate change and global warming **[1]**. Deforestation caused by farming and logging has resulted in the loss of animal habitats **[1]**. The soil is left exposed, resulting in erosion **[1]**, and the land becomes less productive for farming **[1]**, meaning crop yields reduce **[1]**. Economic growth has seen people's incomes rise **[1]**. There are now less people living in poverty [1], meaning they can access healthcare and education **[1]**. They can also afford better housing **[1]**. **[Up to 5 marks]**

Page 114: UK Economic Change

1. **a)** False **[1]** **b)** True **[1]** **c)** True **[1]** **d)** False **[1]** **e)** False **[1]**
2. The number of people employed in farming and mining has declined due to mechanisation **[1]**, so the jobs that people used to do are now done by machines, e.g. combine harvesters in farming **[1]**. Numbers employed in mining have also decreased due to raw material, such as coal, running out **[1]**. Secondary industry has declined due to competition from overseas **[1]**. Cheap labour costs in countries such as Vietnam mean that goods can be made more cheaply than in the UK **[1]**. Some of the production methods used in the UK are outdated **[1]**, meaning UK industries are not competitive compared to those overseas **[1]**. **[Up to 3 marks]**
3. **Any suitable answer, e.g. Bramhall, south of Manchester:** Social changes – population has increased **[1]**; good community spirit with a range of events throughout the year **[1]**; old housing has been modernised **[1]**; local schools are oversubscribed **[1]**; many services are aimed at wealthier people **[1]**; journey times are slow due to congestion **[1]**. Economic changes – new businesses have set up **[1]**; new services have opened due to the growing population **[1]**; jobs have been created in local businesses and services **[1]**; house prices have increased, forcing some people out of the area **[1]**. **[Up to 5 marks]**

Page 114: UK Economic Development

1. UK, USA, Canada, France, Germany, Italy and Japan **[3 if all correct; 2 for up to five correct; 1 for up to three correct]**
2. **Any suitable answers, e.g.** Road building and improvement programmes (including 'smart' motorways) **[1]**; rail network upgrades (including HS2 link) **[1]**; airport expansions (e.g. Heathrow) **[1]**
3. **Any suitable answer, e.g.** The UK trades with lots of countries such as China and the USA **[1]**. Importing and exporting goods and services connects the UK with other countries **[1]**. The UK is a member of the G7 **[1]**, which means it works with other countries to make global decisions **[1]**. Being in the EU has allowed freedom of movement **[1]**, enabling British people to move to other EU nations for work or to retire **[1]**. Also, people from EU member countries, such as Poland, have moved to the UK as economic migrants **[1]**. People in the UK have good access to the internet **[1]**, which enables them to communicate with friends and family in other countries **[1]**. It also means people can buy goods and services from other countries **[1]**. The UK has international transport hubs, such as airports and the Channel Tunnel **[1]**, which enables people to travel abroad for business or tourism **[1]**. **[Up to 5 marks]**

Pages 115–117 Practice Questions

Page 115: Overview of Resources – UK

1. The distance that food is transported from where it is produced **[1]** until it reaches the consumer **[1]**.
2. With increasing affluence, water demand is increasing as new housing is built with more than one bathroom **[1]**, and consumers demand labour-saving devices such as dishwashers and washing machines **[1]**. Industry and agriculture are also large users of water **[1]**.
3. **Any suitable answer, e.g.** Water is transferred from wetter areas to those drier areas with greatest demand **[1]**, using pipelines, aqueducts and rivers **[1]**. Other methods of conservation can be used in the home, such as using showers, low-flush toilets and 'green' appliances **[1]** or water can be recycled using greywater harvesting **[1]**. **[Up to 3 marks]**
4. **Any suitable answer with at least two advantages and two disadvantages, e.g.**
   Advantages: Fracking allows drilling firms to access difficult-to-reach resources of oil and gas **[1]**. It may help to reduce gas prices **[1]**. Electricity can be generated at half the carbon dioxide emissions of coal **[1]**.
   Disadvantages: The use of fracking has prompted environmental concerns **[1]**. It uses huge amounts of water that must be transported to the fracking site at significant environmental cost **[1]**. Cancer-causing chemicals are used and may escape and contaminate groundwater around the fracking site **[1]**. There are worries that the fracking process can cause small earth tremors **[1]**. Environmental campaigners say that fracking is simply distracting energy firms and governments from investing in renewable sources of energy **[1]**. **[Up to 5 marks]**

Page 115: Food 1

1. A lack of food for the population of an area, causing illness and / or death through starvation. **[1]**
2. **Any suitable answers, e.g.** Increase the number of irrigation schemes; restructure farming from subsistence to commercial farming; introduce new high-yielding varieties of staple crops; introduce better storage and transport systems **[3]**
3. **Any suitable answer with two physical and two human factors, e.g.** Physical factors: soils – fertility influences the type of farming; relief – flat land is preferable for arable farming; climate is the most important physical factor **[2]**
   Human factors: cost of land; market inertia – farmers may have farmed in a certain way and may be reluctant to make changes; governments (grants, tax barriers and subsidies); technology, e.g. genetically-modified crops **[2]**
4. **Any suitable answers, e.g.** Climate change is likely to hit the countries of Sub-Saharan Africa the worst **[1]**. Temperatures are likely to rise by more than 2°C **[1]** and this will lead to periods of heatwaves and drought **[1]**, causing desertification **[1]** in some areas and heavy rainfall causing flooding in others **[1]**. This could result in less land available for food production **[1]**. This may lead to increased rates of migration to cities **[1]**. **[Up to 4 marks]**

Page 115: Food 2

1. The application of water to increase the yields of crops, especially in those areas where water is in short supply. **[1]**
2. The 'Blue Revolution' is the growth of aquaculture **[1]** as a very highly productive way of producing food such as aquatic animals and plants in both salt and fresh water **[1]**. Fish farming, such as shrimp or salmon farming **[1]**, and the gathering of seaweed are all methods of aquaculture **[1]**. **[Up to 2 marks]**
3. **Any suitable answers, e.g.** Hydroponics is a method of growing plants using nutrient-rich water **[1]**, often using inputs of ultra-

violet light [1], such as high-value crops like tomatoes in glass-houses [1]. **[Up to 2 marks]**
Aeroponics is a similar way of growing plants in an air or mist environment [1] without the use of soil [1]. It has been used by NASA as a possible way of feeding astronauts on long space journeys [1]. **[Up to 2 marks]**
Biotechnology uses biological processes to develop new products [1] that can help feed the hungry by producing higher crop yields [1]. It can lower the input of agricultural chemicals into crops [1] with less vitamin and nutrient deficiencies [1] and free of allergens and toxins [1]. **[Up to 2 marks]**
Appropriate (or intermediate) technology is technology suited to the social and economic conditions of the people using it [1]. It is environmentally sound [1] and promotes sustainability as it uses locally sourced materials [1]. Examples include digging boreholes to supply water to irrigation schemes [1] or collecting animal dung to convert into methane for gas stoves to use instead of firewood [1]. **[Up to 2 marks]**

4. **Any suitable answers, e.g.** The Gezira Scheme in Sudan [1] is an example of a large-scale irrigation scheme where the flood waters of the Nile [1] are used to grow crops like cotton and wheat [1], but large amounts of water are taken out that affect other irrigation schemes downstream [1]. The cotton is a valuable export for the Sudanese economy [1] but attempts to diversify the type of crops have failed as the country has been affected by civil war [1]. **[Up to 5 marks]**

Page 116: Water 1

1. The inability of a country to ensure sufficient access to quantities of safe water [1] to maintain life, social well-being and economic development [1].
2. **Any two from:** The supply of water is uneven as the distribution of rainfall varies from place to place; arid places have a deficit of water; rising populations increase the demand for water for drinking, bathing, agriculture and industry. **[2]**
3. **Any suitable answer, e.g.** With increasing economic development, the demand for water also rises [1] as domestic use increases through more labour-saving devices like washing machines and dishwashers [1]. Industrial development takes place, leading to a greater demand for water in manufacturing [1] and in electricity generation [1]. Agricultural use of water also increases due to a greater demand for meat and dairy products [1]. **[Up to 4 marks]**

Page 116: Water 2

1. Schemes that move water from one river basin where it is available, to another basin where water is less available. **[1]**
2. Removing salt from either seawater or groundwater to make it drinkable. **[1]**
3. **Any suitable answer, e.g.** The South–North Water Transfer Project in China [1] will enable 44.8 billion cubic metres of water per year [1] to move from the Yangtze River in southern China [1] to the Yellow River Basin in arid northern China [1] through a system of canals and aqueducts [1]. The cost of the project is over $60 billion [1] but is likely to cause water shortages in some parts of China and pollution on some rivers [1]. **[Up to 5 marks]**
4. **Any suitable answer, e.g.** Groundwater makes up nearly 30% of all the world's fresh water [1] and is often used in places where there is not enough water to drink [1]. Groundwater quality is usually very good [1] and needs less treatment than river water to make it safe to drink [1], as the rocks through which the groundwater flows help to remove pollution [1]. Groundwater also responds slowly to changes in rainfall [1], and so it stays available during the summer and during droughts when rivers and streams have dried up [1]. It does not require large costs or technology (as it is often drawn from wells) or expensive reservoirs to store water in before it is used [1]. **[Up to 5 marks]**

Page 116: Energy 1

1. Between 1970 and 2014, the use of oil dropped from 48% to 0% [1], while the use of gas increased from 8% to 30% [1] and coal dropped from 38% to 30% [1].
2. **a)** 19.3% **[1]**
   b) An increase in technology has made the harnessing of these sources possible [1]; investment and subsidies available for installation of wind turbines and solar panels [1]; less reliance on fossil fuels needed to increase energy security [1].
3. North America has large coal resources but its large conventional oil resources are largely exploited [1]; it is now exploiting non-conventional oil and gas reserves but its huge energy consumption often outweighs supplies [1]. Europe is heavily dependent on energy imports [1] as it has declining fossil fuel supply [1].

Page 117: Energy 2

1. **Any suitable answer with examples, e.g.** The capture of energy from existing flows of energy [1], such as sunshine, wind, wave power, flowing water (HEP), biomass and geothermal heat [1].
2. Coal is the dirtiest of fuels (producing 31% of all carbon dioxide) so technology has been developed that removes much of this carbon dioxide before the coal is burned [1]. The reason for this is the impact of greenhouse gases like carbon dioxide on climate change [1]. Laws have been passed to cut down the emissions of coal-fired power stations [1].
3. **Any suitable answer, e.g.** The Muppandal wind farm in Tamil Nadu in southern India [1] has over 3500 wind turbines and is one of the largest in the world [1]. It produces enough electricity to power a million homes [1]. The turbines were purchased by wealthy people, who bought these as an alternative to paying high taxes [1]. As India is a heavy user of coal, installations such as Muppandal reduce the amount of carbon dioxide and the pollution produced [1].
4. **Any suitable answer with four advantages and four disadvantages, e.g.** **[8]**

Advantages	Disadvantages
Geographical limitations: nuclear power plants don't require a lot of space. [1] Nuclear power stations do not contribute to carbon emissions or pollution. [1] Nuclear energy is by far the most concentrated form of energy; a lot of energy is produced from a small amount of fuel. [1] Nuclear power is reliable; it does not depend on the weather. [1] We can control the output from a nuclear power station to fit our needs. [1] Nuclear power produces a small volume of waste. [1]	They have to be located near water for cooling. [1] Disposal of nuclear waste is very expensive: it is radioactive so it has to be disposed of in such a way as it will not pollute the environment. [1] Decommissioning (close down and dismantling) of nuclear power stations is expensive and takes a long time. [1] Nuclear accidents can spread 'radiation-producing particles' over a wide area – this radiation harms the cells of the body which can make humans sick or even cause death. [1] They could be targets for terrorism. [1]

Page 117: Energy 3

1. **Any suitable answer, e.g.** Carbon footprints in HICs tend to be high because of the lifestyles people lead [1], with a reliance on fossil fuels [1], using electrical devices [1] and a diet that relies on food that is imported [1] or uses high levels of inputs derived from fossil fuels [1]. **[Up to 4 marks]**
2. **Any suitable answer, e.g.** In LICs, carbon footprints tend to be low as lifestyles are more sustainable [1] as less fossil fuel is used [1] and food tends to be locally produced [1]. **[Up to 2 marks]**
3. **Any suitable answer, e.g.** Energy can be conserved through the design of homes by the use of insulation in loft spaces and walls [1], double-glazed windows [1] and larger windows in south-facing walls [1]. **[Up to 2 marks]**
4. **Any suitable answer, e.g.** Using energy-efficient devices like kettles that only boil one cup of water [1], along with turning off appliances when not in use (e.g. not using standby buttons), help to conserve energy in the home, [1]. Other methods include the use of LED light bulbs [1], washing machines that wash clothes at 30°C or lower [1] and thermostats on central heating systems that can be controlled using a mobile phone [1]. **[Up to 4 marks]**

Page 117: Fieldwork

For all questions, you need to know the titles and be able to describe the location for both of your enquiries. The answers given below are suggestions for things you may consider depending on the nature of your enquiries.

1. **a) and b)**
   - Secondary data: cost-benefit data; comparing old and contemporary maps, aerial images, photographs; virtual fieldwork (e.g. Street View; geography.org.uk).
   - Primary data: environmental quality survey (to produce

environmental quality index); land use mapping; questionnaires; oral histories; field sketches; photographs.

a) Specific to physical environment
Rivers / Coast:
- Secondary data: Government information such as flood-risk maps from Environment Agency; cost-benefit data; comparing old and contemporary maps, aerial images, photographs.
- Primary data collection methods for depth, width, wetted perimeter, velocity, gradient, sediment size and shape (roughness) using equipment such as a tape measure, range poles, chain, metre ruler, clinometer, pantometer, hydro-prop with impeller, flow meter, floating object (e.g. table tennis ball, cork, orange, stick), stopwatch, callipers. Physical maps (e.g. geology, drainage, floodplain zones, coastal management zones). As appropriate, profiles or transects might be produced. **[Up to 1 mark]**

b) Specific to human environment
Urban / Rural:
- Secondary data: Government information such as census data from the Office of National Statistics; census data for age profile, unemployment, qualifications, housing tenure, house price or rental survey, Index of Multiple Deprivation (composite indicator), rural deprivation. Human geography maps (e.g. infrastructure maps; Goad maps; neighbourhood zones; planning maps).
- Primary data collection methods for traffic counts; car age or origin; pedestrian counts; shopping quality (e.g. independents vs. chains); rate of turnover for retail premises; footfall; index of decay; building heights; noise levels or air quality (could use a smartphone app). **[Up to 1 mark]**

2. For both environments:
   - Ease of access physically and legally (e.g. land ownership); distance from school and transport availability; size of area or ward in which representative data can be gathered in time available.
   - Safety: risk assessment and measures to reduce risk; links to hypotheses about the area; ability to make comparisons between contrasting places or to observe changes since earlier investigations took place.

   Specific to physical environment
   - Rivers / Coast: Manageable size, e.g. small area in which students can walk up and down and, if necessary, be near water or steep slopes or cliffs easily and safely. **[Up to 1 mark]**

   Specific to human environment
   - Urban / Rural: Manageable size, e.g. small area(s) of a settlement that can be accessed easily and safely on foot; traffic. **[Up to 1 mark]**
3. **Briefly explain the data collection method. 'Justify' means that you need to explain why it was effective and how it promoted your investigation. State whether the data was quantitative or qualitative; explain the advantages of such data. Describe and explain sampling techniques you used to ensure the reliability of your data, e.g. random, systematic, stratified. If you used GIS, explain why such georeferenced data is a useful technique. State how the data contributed to answering a key question or testing a hypothesis. [Up to 3 marks]**
4. **Briefly explain the limitations of a data collection method (e.g. why errors may have occurred) and how it might be improved. A good answer should consider questions such as these:** If the sampling technique was a limiting factor, what alternative sampling technique might be used? If the sample size was unrepresentative, could more samples be taken? If the reliability of data is an issue due to when it was gathered, would the findings be more reliable if the investigation was repeated at a different time of day / different day of the week / different time of the year? If the reliability of data is an issue due to inconsistent data gathering (e.g. students not using equipment or techniques in the same way), would more training or practice be helpful? **[Up to 3 marks]**

Pages 118–120 Review Questions

Page 118: Overview of Resources – UK

1. A carbon footprint measures the total greenhouse gas emissions caused directly and indirectly **[1]** and is measured in tonnes of carbon dioxide **[1]**.
2. As the UK relies on fossil fuels and domestic supplies of coal, gas and oil are running out **[1]**, it means it has to import more from other countries **[1]**.
3. **Any suitable answer, e.g.** There are likely to be conflicts between the gas companies and local people in the areas where the gas is drilled **[1]**. This may be due to the impact of traffic **[1]**, pollution of water supplies **[1]** or the possibility of earth tremors **[1]**. There may also be conflict between gas companies and environmentalists **[1]**. **[Up to 3 marks]**
4. **Any suitable answer, e.g.** In the 1960s, most of the UK's electricity was generated using coal, oil and a small amount of nuclear **[1]**. Gas was also made from coal **[1]**. Now around half of the UK's electricity is made from gas **[1]** and the rest from coal, nuclear and renewables **[1]**. More fuel-efficient cars mean that only the same amount of oil is used for transport despite there being three times as many cars **[1]**. **[Up to 4 marks]**

Page 118: Food 1

1. When all people at all times have access to sufficient, safe, affordable nutritious food **[1]** to maintain a healthy and active life **[1]**.
2. **Any two from:** Climatic factors (too hot, too cold, too dry, too wet); water availability (for irrigation); soil type / fertility. **[2]**
3. **Any two from:** Population size (the ability of the country to support its people); insufficient farming skills; insufficient financial investment into a country's agricultural industry. **[2]**
4. **Any four from:** Farmers are poor and are stuck in the cycle of poverty; lack of investment in agriculture due to poverty; climatic factors including climate change and drought; civil war; price fluctuations leading to unstable markets; food wastage; poor storage and distribution networks **[4]**

Page 118: Food 2

1. **Any suitable answer, e.g.** Organic farming is a sustainable method of food production **[1]** that does not use synthetic pesticides or fertilisers **[1]**, instead using organic products to fertilise **[1]** (plant and animal waste) and biological solutions for pest and disease control **[1]**. **[Up to 2 marks]**
2. **Any suitable answer, e.g.** Reducing food waste and losses in both our homes and in shops could save UK consumers up to £2.4 billion a year. The average UK family throws away enough food for around six meals a week **[1]**. Buying less food in the supermarket and using goods before they go 'off' would lead to a more sustainable lifestyle **[1]**.
3. **Any suitable answer, e.g.** The Green Revolution is an initiative that has increased agricultural production **[1]**, particularly in LICs since the late 1960s **[1]**. The Green Revolution saved around a billion people from starvation **[1]** and involved the development of high-yielding varieties of cereals **[1]** (rice and wheat) and new farming techniques **[1]** including the use of irrigation **[1]**, synthetic fertilisers **[1]** and pesticides **[1]**. **[Up to 4 marks]**
4. **Any suitable answer, e.g.** The Organopónicos in Havana, Cuba, is a good example of a local scheme to increase food security among the urban poor **[1]**. It was introduced after the collapse of the Soviet Union, when Cuba lost its ability to produce enough food for its population **[1]**. It is an example of urban farming that is a sustainable practice, allowing urban dwellers to produce enough food for their families **[1]**, and can include a variety of activities such as vegetable and fruit growing **[1]**. Organopónicos take up 3.4% of urban land country-wide, and 8% of land in Havana **[1]**. They produce over 3 million tonnes of organic food **[1]** and calorie food intake is back at 2600 calories a day after the threat of hunger was a possibility **[1]**. **[Up to 5 marks]**

Page 118: Water 1

1. When there is not enough water to meet all demands as a result of physical conditions. **[1]**
2. A lack of investment in water systems or insufficient human capacity to meet the demand for water in areas where the population cannot afford to use an adequate source of water. **[2]**
3. **Any suitable answer, e.g.** LICs can improve water security by reducing wastage and evaporation **[1]**, transferring water from regions of surplus to regions of deficit **[1]** and encouraging small-scale schemes at a local level **[1]**. **[Up to 2 marks]**
4. **Any suitable answer, e.g.** Climate change is altering patterns of rainfall around the world **[1]**, causing shortages and droughts in some areas **[1]**, leading to desertification **[1]**. In the future, two-thirds of the world's population may face water shortages **[1]**. **[Up to 3 marks]**

Page 119: Water 2

1. Water held underground in the soil or in spaces in rock. **[1]**
2. Diverting supplies from one river system to another river **[1]**; the building of dams and reservoirs increases the supply of water **[1]**; a number of countries have opened desalination plants **[1]**.

3. Groundwater is often found in porous rocks deep underground in reservoirs called aquifers. Groundwater is relied on in many developing countries, especially in Africa, because it can often be found close to villages **[1]**. Aquifers are often slow to recharge (fill up) **[1]**, so may not always be sustainable, and groundwater supplies can become contaminated **[1]**.
4. **Any suitable answer, e.g.** The Agra clean water project, India. Agra is a city of 1.3 million people **[1]**. The city's water supply is dependent on the Yamuna River, which provides a limited supply of polluted, undrinkable water **[1]**. Those who can afford it purchase bottled water or household filters **[1]**, while people living in one of the city's 432 slums generally either tap groundwater supplies or depend on private tankers, which bring in water from outside of the city **[1]**. The project covers two areas of slums in Agra **[1]**. They are not connected to the water network and the groundwater is highly polluted by local industry **[1]**. The Agra clean water project will combine water testing and an education programme **[1]** to revive traditional knowledge and systems of rainwater harvesting and water conservation **[1]** to ensure that around 2500 people have their water supplies improved **[1]**. **[Up to 5 marks]**

Page 119: Energy 1

1. The uninterrupted availability of energy sources at an affordable price. **[1]**
2. A lack of fuel resources available **[1]** and a reliance on imported fossil fuels **[1]**.
3. **Any suitable answer, e.g.** Agriculture uses oil products to power farm machinery **[1]**, for transport of goods and livestock **[1]**, and in agricultural chemicals such as fertilisers and pesticides **[1]**. In recent years, agricultural goods like barley, maize and sugar cane have been used to make biofuels **[1]** that are used as a substitute for oil-based fuels **[1]**. Rising prices for oil mean a higher price for biofuels and agricultural chemicals **[1]**, making food more expensive **[1]**.
4. **Any suitable answer, e.g.** Demand for fossil fuels, especially oil and gas, may mean that producers have to exploit areas that are challenging **[1]** such as in deep oceans **[1]**, polar regions **[1]** and other remote, difficult and environmentally sensitive areas. The development of technology has enabled this **[1]**.
(Marks are also available for examples, such as, in the 1960s British companies searched for oil and gas in rocks underneath the North Sea **[1]**, and, in Alaska, exploitation of oil sources has taken place in environmentally sensitive and remote areas close to the Arctic Ocean **[1]**.) **[Up to 5 marks]**

Page 119: Energy 2

1. Fuels produced directly or indirectly from organic material including plant materials and animal waste **[1]**.
2. **Any suitable answer, e.g.** Countries that are LICs may not be able to afford to build them **[1]**. If they receive low amounts of rainfall, this may cause water shortages **[1]**. If there is only flat land, valleys cannot be made into dams successfully **[1]**. Dams cause environmental damage to rivers **[1]**. **[Up to 3 marks]**
3. Geothermal is a sustainable source of energy using heat from the Earth **[1]**. It is mainly exploited in areas of volcanic activity, such as Iceland **[1]**. The steam produced is used to power turbines to create electricity **[1]**.
4. **Any suitable answer with two advantages and two disadvantages, e.g.**
Advantages: wind is a renewable energy source and there are no fuel costs **[1]**; no harmful polluting gases are produced **[1]**.
Disadvantages: wind farms are noisy and may spoil the view for people living near them **[1]**; the amount of electricity generated depends on the strength of the wind / if there is no wind, there is no electricity **[1]**.

Page 119: Energy 3

1. Safeguarding **[1]** and cutting down on the use **[1]** of precious supplies of energy.
2. **Any suitable answer, e.g.** Energy can be conserved in buildings with methods such as the use of insulation in loft spaces **[1]** and walls **[1]**, double-glazed windows **[1]** and larger windows in south-facing walls **[1]**. **[Up to 3 marks]**
3. **Any suitable answer, e.g.** Transport in cities can be made more sustainable by encouraging people to use bicycles rather than cars **[1]**, providing bike lanes **[1]** and offering bikes for hire **[1]**. Other examples include the congestion charge in central London **[1]**, park-and-ride schemes **[1]**, tram networks **[1]** and hybrid buses **[1]**. **[Up to 5 marks]**
4. Technology that enables coal-fired power stations to be efficient in the use of fossil fuels **[1]** and reduce the amount of greenhouse gases produced **[1]**.

Page 120: Fieldwork

For all questions, you will need to know the titles and be able to describe the location for both of your enquiries. The answers given below are suggestions for things you may consider depending on the nature of your enquiries.

1.
- For quantitative data: scattergraphs, line charts, pie charts, bar charts, histograms with equal class intervals, particular types of bar charts such as population pyramids, divided and cumulative bar and line charts, pictograms, proportional symbols, choropleth maps, isolines, heat maps, dot maps, desire lines and flow-lines. You could also include tools for statistical analysis such as lines of best fit; measures that show the strength of correlation, such as R coefficients; Spearman's Rank correlation; measures of central tendency: median, mean, mode and modal class; measures of spread and cumulative frequency: range, quartiles and inter-quartile range, dispersion graphs; percentages: percentage increase or decrease, percentiles; relationships in bivariate data: trend lines through scatter plots, lines of best fit, positive and negative correlation, strength of correlation; predictions and trends: interpolation, extrapolation; limitations and weaknesses in selective statistical presentation of data; geo-spatial (geolocated or georeferenced) data presented in a geographical information system (GIS) framework; GIS can also be used to analyse spatial data.
- For qualitative data: annotated images, diagrams or field sketches, overlays using GIS or tracing paper, tables comparing quotes or opinions.
[Up to 3 marks]

2. For the technique used, briefly explain what the presentation technique is and how it shows or enhances the information provided by the data. **[Up to 4 marks]**
3. **Structure your answer, considering the following:**
Potential risks and likelihood: slipping (high), clothing becoming wet, footwear leaking (high), hypothermia (medium – depends on weather), drowning (water not deep so low risk), allergies, traffic / wildlife (low), local conditions, e.g. stability of slopes.
Who may be affected: students, teachers and other accompanying adults, members of the public. **[Up to 4 marks]**
4. **A good answer should consider the following:**
What are the potential risks? Who may be affected by the risks? What is the level or likelihood of the risk (high, medium, low)? What can be done to reduce / manage the risk? If these measures are taken, how likely is the risk to happen then?
Here is an example of a student's notes for writing a risk assessment:
Potential risk: water in river, nearby streams, traffic near coach;
People affected by the risks: students / staff falling in the water, becoming dangerously cold, slipping or falling on rocks;
Level / likelihood: medium for most, traffic – low;
Risk reduction / management: appropriate clothing and footwear, discussion and advice about risks beforehand, reminders at key times;
Impact of measures: reduced likelihood; if something happens, everyone is prepared and knows what to do. **[Up to 4 marks]**

Pages 121–127 Mixed Questions

1. **Any suitable answer, e.g.** The USA and France are HICs **[1]** and use large amounts of water for agriculture, industry and domestic purposes **[1]**. China and India are NEEs **[1]** and their water usage is increasing due to industrialisation **[1]** and increasing domestic demand **[1]**. Mali and Egypt are both arid countries **[1]** but Mali is a LIC whereas Egypt is a NEE **[1]** and uses lots of water for irrigation purposes, having plentiful supplies (River Nile) **[1]**. **[Up to 5 marks]**
2. An extreme natural event (e.g. earthquake, volcano, tropical storm) **[1]** that causes loss of life and / or severe damage to property or severe disruption to human activities **[1]**.
3. **Any suitable impacts occurring immediately, e.g.** Falling glass; falling masonry; loss of power supplies. **[2]**
4. **a)** False **[1]** **b)** True **[1]** **c)** True **[1]** **d)** True **[1]** **e)** True **[1]**

5. The Saffir–Simpson scale classifies a tropical storm into one of five categories [1] based on its sustained wind speed [1].
6. The UK's location in the mid-latitudes on the coast of north-west Europe beside the Atlantic means that there is frequent conflict between cold and warm air masses [1] creating fronts, which produce precipitation [1].
7. hockey stick [1]
8. **Any suitable answer, e.g.** Too many beetles born in one year means more food for the birds that prey on them, so the beetle population is reduced to a sustainable level. [2]
9. **Any suitable answer which gives the idea that replenishment is needed as there are always processes at work which remove surface debris from beaches, e.g.** Beaches can be subject to longshore drift, which removes material [1] along the shore. Material is taken offshore by destructive waves [1]. Unless more material is added from weathered and eroded cliffs [1] or longshore drift [1], the beach will disappear. **[Up to 3 marks – maximum of 2 marks if only removal process(es) described]**
10. Taiga [1]
11. Salinisation [1]
12. Ships increased in size [1]
13. **Any suitable answers, e.g.** Tree-labelling schemes [1]; debt reduction [1]
14. Causes: Intense low pressure in a tropical storm creates a dome of seawater, especially around the eye of the storm. As this moves across land, floods occur. [2]
Potential impacts: coastal floods; infrastructure damaged; drownings; if sewage becomes mixed with floodwater, disease can spread. [2]
15. Rural to urban migration of agricultural workers [1]
16. Frost-shattering [1]
17. **Any suitable answer, e.g.** Steep slopes make roads or railways difficult / impossible to build [1]. Troughs form good routeways [1] but building may need to be on the sides [1] to guard against flooding on the flat floor [1]. **[Up to 2 marks]**
18. **Any suitable impacts occurring later on, e.g.** Loss of jobs; polluted water supplies; crime (looting) [2]
19. **Any suitable answer, e.g.** Installing solar panels [1]; using energy-efficient devices like LED light bulbs [1] and plastic kettles [1]; turning off appliances when not in use (rather than using standby buttons) [1]
20. **Any suitable answer, e.g.** The use of showers and low-flush toilets [1] along with 'green' appliances that use little water [1]. Greywater harvesting [1] is a way of conserving water involving the recycling of water used in baths and showers [1] as well as from rainwater from roofs to use for flushing toilets and other non-drinking purposes [1]. **[Up to 4 marks]**
21. **Any suitable answer, e.g.** Costs involved in purchasing the technology; fuel; availability of replacement parts and maintenance [2]
22. **Any suitable answer with three points (at least one action and one result), e.g.** Particles held in the water [1] are used as tools [1] to scrape / scour the channel bed and banks [1]. The channel is cut downwards [1] and can be widened [1], especially on bends. **[Up to 3 marks]**
23. a) Notch [1]
b) **Most likely answers are salt crystal growth or frost-shattering (or plant root and / or animal burrowing):**
Salt crystal: salt grains from sea spray / seawater dries in cracks / joints [1] and expands when wet again to put pressure on rocks [1].
Frost-shattering: water in cracks / joints expands on freezing [1] and puts sufficient pressure on rocks to break them [1].
Plant roots / animal burrowing: plant roots and / or animal movement widens cracks, joints and bedding planes [1], causing fragments of rock to break away [1].
c) Rockfall (or 'fall') [1]
24. **Any suitable answer, e.g.** Calorie intake varies between rich and poor countries. People in the USA and Europe, with over 3000 per day [1], consume the most calories per capita on average [1] and this is much higher than the recommended daily calorie intake of 2500 for men and 2000 for women [1], leading to obesity in these countries. Sub-Saharan Africa has many countries experiencing below average per capita calorie intake, with averages of 2300 [1] calories per day for the region, while South Asia has an average of around 2700 [1]. **[Up to 3 marks]**
25. A greenfield site is one that has never been used for development before [1]; a brownfield site has had a previous (now redundant) industrial use [1].
26. Previous tributary river valleys have lost their confluence [1] due to erosion of the spurs into truncated spurs [1], so water tumbles over the edge.
27. **Any suitable answer – key elements are:** line of most efficient flow; erosion on outer bend(s); deposition on inner bend(s); pushing out of a bend into a definite meander.

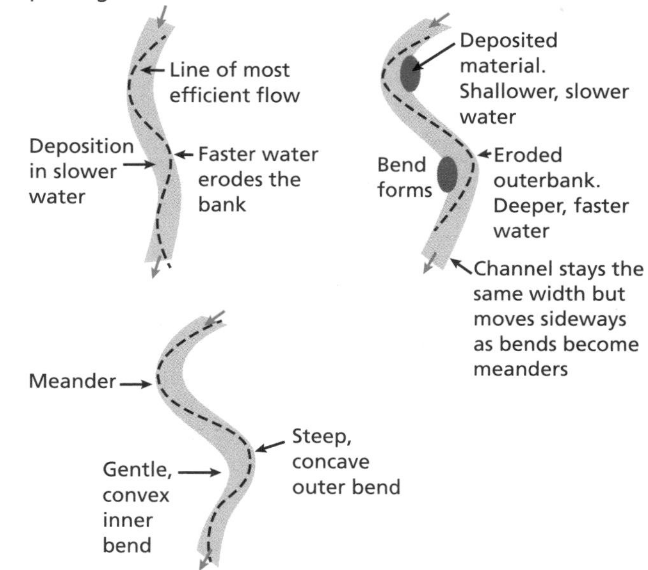

[1 for diagrams; 3 for annotations / labels]
28. 60% [1]
29. **Any suitable answers, e.g.** Lack of affordable housing; loss of communities; loss of jobs; traffic congestion; loss of local services [3]
30. Life expectancy, literacy, and income **[2 if all correct; 1 if two correct]**
31. Greater use of nuclear [1] and renewables [1] to generate electricity.
32. **Any suitable answer, e.g.** Ethical consumerism is the purchase of sustainable goods [1], such as sources of food that are locally grown [1], and only eating fruit and vegetables that are in season [1]. **[Up to 2 marks]**
33. **Any suitable answer with four linked points (it should mention at least one specific erosion process and debris removal, and there may be reference to back-to-back caves), e.g.** Waves weaken the back of a cave using both hydraulic action [1] and abrasion [1], causing the rock to break away [1] and remove loose material [1]. At high tides, the roof of a cave will be eroded [1]. Eventually, the back of the cave breaks through to the other side of its headland [1].
34. **Any suitable answer, e.g.** Carbon footprints in HICs tend to be higher [1] due to the lifestyles people lead, as they have a greater reliance on fossil fuels [1] and they tend to use electrical devices in the home and at work [1]. People in HICs have a diet that tends to rely on food that is imported [1] or is grown using high levels of inputs derived from fossil fuels [1]. In LICs, carbon footprints are generally lower [1]. Lifestyles are more sustainable because lower amounts of fossil fuels are used [1] and food tends to be locally produced [1]. **[Up to 6 marks]**
35. **Answers will vary but should consider these points:**
Reliability of conclusions: For the topic of your enquiry, you need to make judgements about how close your conclusions are to actual changes happening.
Comments on reliability may be affected by limitations of equipment used; equipment operator errors; choice of data collected; methods of data collection; design methodology / sampling methodology, e.g. number and suitability of sample sites (spatial) and time of day or year (temporal).
Use these ideas to reach an overall judgement about the reliability of your conclusions. Also evaluate the extent to which your findings would be repeated if the investigation was undertaken at different places (spatial) or times (temporal).
[9 + 3 SPaG]

Glossary and Index

3 Ps: Prediction; Protection; Planning measures a community or country can take to improve their resilience to a hazard such as a tropical storm 9, 17
abrasion erosion process; use of material carried in water (sea, river) to scour and scrape 44–45, 51, 57
active volcano a volcano with at least one eruption in the last 10 000 years 12
adaptation measures taken to reduce the negative impacts, or take advantage of the positive impacts of, climate change 21, 30–31, 36–37, 107
aesthetic a sense of beauty or attractiveness 48
air masses large bodies of air with the same temperature throughout 18–19, 28
alluvium material deposited by rivers 59
altitude height above sea level 18, 48
angular sharp-edged shape of rocks, not rounded at all 44
anthropogenic caused or produced by humans 21
appropriate technology science or technology suitable for the area 35, 79, 96
aquifer an underground layer of rock that bears water 35, 100–101
arable the farming of crops 48–49
arch erosion feature; headland that has been cut through away from the seaward end to form a bridge shape 52
arête steep-sided, narrow, rocky ridge formed between corries 46–47
arid region an area with a severe lack of available water 98–99
asymmetrical not identical either side of the middle; in rivers, a channel almost always has one bank steeper than the other 6, 47
atmospheric circulation the movement of air around the Earth in cells, transferring and redistributing energy 14, 28
atmospheric pressure the pressure exerted by the weight of the atmosphere 28
attrition process in which particles get smaller and rounder as they crash into each other in water (sea, river) 51, 57
backwash return wave motion to the sea 50–51
bar strip of deposited sediment parallel to the coast 53
beach deposition feature; loose material, usually sand or shingle 53
bedload material on the river channel floor; generally the biggest particles and most difficult to move 57
biofuels fuels produced from renewable resources such as vegetable oils 103, 104
biomass renewable material from living things 104
blocking high slow-moving anticyclone (high-pressure system) 18
brownfield site land previously used for industrial purposes 73, 85
buttress roots a shallow root system ensuring the quick take up of nutrients 30
carbon capture (carbon sequestration) the removal of carbon at emission source and storage of it deep underground 21
carbon footprint the amount of carbon dioxide released as a result of an organisation's/individual's activity 92, 106
carbon neutral taking action to remove as much carbon dioxide from the atmosphere as is put into it 74
carnivore an organism, such as an owl, that eats other animals 26–27
channel the groove in which water flows 56–57, 58–59, 60–61
channel catch rainwater falling directly into the river 60–61
choropleth map uses shading to indicate average values 7
clean coal technology seeking to reduce harmful coal-based emissions by using technology 104, 107
climate average weather pattern over time; expected conditions 4, 18, 29, 30–31, 48
collision zone a plate margin where two tectonic plates crash into each other 9
colonialism where one country goes into another country and claims they are in power, often exploiting people and resources 78
commercial farming crops are produced on a large-scale for profit 34
composite volcano cone-shaped volcano made up of many layers 12
confluence where any two rivers meet 46, 56–57
conservative margin formed when plates slide past each other 8
constructive margin formed when plates move apart 8
constructive wave low energy waves which add material to a beach; more effective swash than backwash 50
consumer a species that eats another species 26–27
continental coming from a nearby continent 18
contours lines along which the height is the same 5, 7
core the central region of the Earth 8–9
Coriolis force (Coriolis effect) the result of Earth's rotation on weather patterns 15
corrie curved rock hollow in a mountainside with a steep back and ice-carved base; may contain a lake 46–47, 48
counterurbanisation where people move from urban areas to rural areas 83
critical thinking the ability to interpret, analyse and evaluate ideas and arguments 110–111
crust the topmost layer of the Earth, made up of tectonic plates 8
cumec cubic metres per second; the measurement of discharge in a river 60
cyclone a tropical storm in the Indian Ocean and South Pacific, e.g. India 14
dairy the production of milk 48
debris loose material that has been weathered or worn away from a landscape and can be carried away or deposited 44–45, 46–47, 52
decompose to break down 50
decomposer an organism, such as fungus, that breaks down the dead remains of another organism 27
de-industrialisation a decline in the number of workers or the output in manufacturing industry 82
Demographic Transition Model shows birth rate, death rate and population change over time 77
dense pattern or distribution is dense if it shows a high concentration in one area, e.g. dense drainage pattern; dense population 4
deposition material being left behind in response to a loss of energy 45, 47, 50, 51, 57, 58–59
desalination removing minerals from saltwater to render the water suitable for human use 100
desertification the process by which previously fertile land becomes desert 34–35
destructive margin a plate boundary that occurs when plates move together 8
destructive wave high energy waves which remove material from a beach; backwash is more effective than swash 50
dispersed a type of settlement pattern: separate and scattered 4
diversification use of farm land or buildings for non-farming activities 49
dormant volcano an active volcano that is not currently erupting 12
downcutting vertical erosion of a channel, making it deeper 57
drainage basin water from an area that feeds into a river; also known as a river basin or a catchment area 56
dredging scraping material out of a channel to make it bigger 61
drought a period of below average precipitation 18–19, 95
drumlin elongated glacial deposit with a blunt upstream side (stoss end) and tapered downstream side 47
easting line one of the vertical lines crossing an OS map from top to bottom; they are called eastings as the numbers increase in an easterly direction 4
economic development improvement in living standards by creation of jobs 32, 98, 103
economic leakage when money that is generated in a country leaves a country; this happens when a transnational corporation sends money back to its home country 80
economic water scarcity the population does not have the monetary means to access available water 98
ecosystem a community of interdependent living and non-living elements that create a particular environment 26–27, 28–29, 30–31
ecotourism responsible travel to areas that sustains the well-being of the local population and environment 33
elevation profile shows how far above sea level certain features are 5
embankment artificial bank built alongside a channel; can be of natural materials or concrete 61
energy consumption amount of energy used by individuals or groups 102–103
energy insecurity without access to a secure and affordable energy supply 102–103
energy poverty unable to heat or provide other energy services to homes 102–103
energy security uninterrupted availability of energy sources at an affordable price 93, 102
en-glacial (moraine) debris inside the ice 45
enterprise zones areas which provide incentives to attract business investment 83
entrained material being taken into an agent of transport and erosion, such as a glacier or river 56
environmental relating to the natural world and the impact of human activity on its condition 48, 72, 73
epicentre the point on the Earth directly above the focus of an earthquake 10–11
erosion wearing away of the landscape due to the movement of ice, water or wind 44–45, 51, 52, 56–57
erratic a rock or boulder that differs from the surrounding rock 47
estuary tidal river mouth where saltwater from the sea meets freshwater 53, 59
ethical consumerism buying only products or services that are produced in a way that does not harm the environment 97
extinct volcano a volcano that has not shown an eruption for at least 10 000 years, and is not expected to erupt again in the future 12
extrapolation predicting future outcomes based on known facts 7
extreme temperatures excessive heat or cold 31, 34
extreme tourism travel to remote or unsettled areas which can be dangerous 37
eye of the storm if a tropical storm becomes 'cyclonic', it spins so fast that the air around the centre forms a vortex which has an eye 20–40 miles (30–65 km) wide 15
fair trade farmers paid a fair price for their goods 79
famine extreme scarcity of food 93
faulting breaks in rocks; can be as small as a hairline or huge 58
favela an area of makeshift housing in or near a city in Brazil 70–71
fetch distance over which a wave has travelled 50
flashy a river that responds quickly to rainstorms 60
flocculate to cause material to join together, gain weight and sink to the bottom of a river/sea 59
flood a river going over its bank 48, 57, 59, 60–61
floodplain low-lying area made of deposited material to sides of a river channel; flood waters contribute to the formation 57, 59
fluvial the processes associated with rivers and streams 56
flood management measures which are taken to alleviate the negative impacts of flooding on people and their property 6, 60–61
fodder coarse food for livestock, e.g. hay, straw 48
food chain simple set of connections showing how a small group of organisms are linked; who eats what 26
food insecurity being without access to enough food 94–95, 96–97
food miles the journey and fuel used from food producer to consumer 75, 92
food security having reliable access to enough food 94–95, 96–97
food web diagram showing how larger groups of organisms are interlinked 26–27
fossil fuel sources of energy formed from the remains of organisms buried millions of years ago; coal, oil and gas 102–103, 104–105, 106–107
fracking drilling the earth and injecting high pressure liquid to extract gas 93

freeze-thaw breaking down rocks by frost shattering, that is water entering cracks, expanding when frozen and so putting pressure on the rock and weakening it 44, 46
frequency regularity of an event such as a tropical storm 7, 15
friction force between surfaces trying to pass each other; in a river, there is friction between the water and channel 8, 58
front a boundary separating two masses of air of different densities 18
gabions bundles of rock in a wire mesh cage; used to protect slopes, especially at the coast, on river banks, and roads 54
geographical enquiry investigation linking your study in school/college with fieldwork carried out away from school/college 108
geology the science of Earth's origins, including rocks and minerals 50, 60
geothermal energy using heat from deep within the ground (such as near volcanoes) to obtain energy 9, 105
GIS systems designed to capture, store, manipulate, analyse, manage, and present all types of spatial or geographical data; GIS data is geolocated or georeferenced using widely recognised locational methods such as latitude and longitude 5, 7, 109, 110
glacials cold periods in ice ages when glaciers and ice sheets grow 20
glacial trough deep, steep-sided valley formed by glaciers 46–47
glacier slow moving stream of ice formed from accumulation of snow 44–45, 46
globalisation process whereby the world is becoming increasingly interconnected thanks to mass communications 82
gorge steep-sided section of a valley; usually rocky 58
gradient steepness of a slope 53
greenfield site a plot of land that has not been used before for building development 85
greenhouse effect greenhouse gases stop heat escaping from the Earth into space; an enhanced greenhouse effect can lead to climate change and global warming 20–21, 104
greenhouse gas gas in the atmosphere which absorbs and emits heat (infrared) radiation, such as carbon dioxide, methane and water vapour 20–21
greywater waste water from the household that can be recycled 98, 101
ground moraine thick layer of clay containing rock fragments from under a glacier; also known as till 47
groundwater water held underground in the soil or rock 98–99, 100–101
groundwater flow slow, underground, horizontal movement of water towards the channel 60
growing season a period of the year when crops grow successfully 48
groynes man-made structures at right angles to the coast to stop material travelling along the beach by the action of longshore drift 54–55
Gulf Stream a powerful warm ocean current 18
hanging valley tributary valley along the side of a glaciated valley 46
hard river engineering range of measures used to manage river flow and floods which use artificial structures made from materials such as concrete, stone or steel, e.g. wing-dykes; dams; weirs; often more expensive than soft engineering and considered to be less economically, socially and environmentally sustainable 6, 61
headward towards the source of a river 56
helicoidal like a flattened curve; water takes this motion at meander outer bends, hitting the bank high then moving down and outwards towards the next inner bend 58–59
herbivore an organism such as a sheep that eats only plants 26–27
HIC higher income country 10–11, 68, 76–77
histogram data grouped into ranges and plotted as bars 7
hot spot where strong upward currents in the magma reach close to the surface at points on the crust away from plate margins; volcanic activity is common here 9
Human Development Index (HDI) a measure of development taking account of several development indicators 76
hunter-gatherers tribespeople who live by hunting and harvesting wild food 32
hurricane a tropical storm with heavy rain and strong winds 14–15
hybrid vehicle one which uses two or more types of power, such as internal combustion engine plus electric motor 107
hydraulic action erosion process; action of moving water in rivers; air trapped in cracks etc. by waves exerts great pressure 51, 56–57, 58
hydroelectric power (HEP) producing electricity using water power 32, 34, 105
hydrograph diagram showing the amount of water in a particular river over time; can be produced for minutes, hours, days, months, years 60
hydro-meteorological hazards weather hazards linked to the water cycle 18
hypothesis a statement or theory that can be tested 108–109
impacts the economic, social and environmental effects of an action or decision 19, 21, 111
impermeable surface that does not allow water to pass through 6, 48–49, 60–61
industry economic activity not usually associated with agriculture 48, 79, 84
inertia when people and/or organisations do not take action when faced with the risk of a hazard 17
infrastructure facilities and supply lines which make modern life possible, such as roads, railways, air and sea ports, water pipes, sewage pipes, electricity cables, telecommunications links, computer networks and Internet access 17, 34–35, 84
intensity severity or strength of an event such as a tropical storm 15
intercept get in the way of; anything that stops rainwater reaching the ground, permanently or temporarily 61
interest group organisation of people with a common cause about which they try to influence policies or decisions, without seeking political control 110–111
interglacials warmer periods in ice ages 20
interlocking spurs fingers of land around which rivers flow; they obstruct the view up or down the valley 58
intermediate technology technology appropriate for the developing needs of the community 79
interpolation predicting values from a limited number of sample data points 7
interrelationship two-way links between different elements in an ecosystem 26
IPCC the Intergovernmental Panel on Climate Change 20–21
irrigation artificial addition of water to land to help crops grow 30–31, 34–35, 96–97
isobars lines on a weather map joining areas of the same air pressure 6
isohyets lines on a weather map joining areas of the same rainfall 6
isolines lines drawn to link different places that share a common value 5, 6–7
key players people or groups with an interest and a significant degree of control about an issue or in the development of a decision or policy 110–111
lag time difference in time between peak rainfall and peak discharge 60
lahar a mud flow consisting of volcanic ash and water that runs down the slopes of a volcano, sometimes burying settlements in its path 12–13
landlocked surrounded by land and having no navigable route to the sea 78
landward facing away from the sea 55
lateral (moraine) material carried at the side of a glacier and deposited along the side of a valley 45, 47
latitude latitude lines provide a measure of how far places are north or south of the Equator; latitude lines are all parallel to the Equator 4
latosol soil found under tropical rainforests 30
leaf litter dead plant material on the ground surface 30–31
levee ridge of deposited material alongside a channel; results from floods; artificial levees can be made as flood defences 59
LIC lower income country 11, 68, 76–77
limestone sedimentary carbonate rock made from the bodies of sea creatures 57
limitations disadvantages or causes of potential errors or the collection of unreliable or unrepresentative data 109
linear pattern or distribution is linear if there is a concentration of something in a line or ribbon, e.g. a linear settlement pattern 4
load material carried by an agent of erosion and transport, e.g. river, glacier 57
longitude longitude lines provide a measure of how far places are east or west from the Prime Meridian (sometimes called the Greenwich Meridian); longitude lines run between the north and south poles at right angles to the latitude lines 4
longshore drift movement of material along the shore due to wave action 51, 53, 54–55
long-term aid providing help for education and skills for development 79
lyme grass salt-loving plant; one of the first to grow on sand dunes 55
management the forecasting, planning, organising and decision-making used by people to improve people's lives 111
mantle the region within the Earth between the core and the crust 8–9
maritime related to the sea 18
marram grass tough, long rooted, wide bladed grass that lives in sand and holds the grains together 55
mass movement downslope movement of surface debris under gravity 50–51
mass tourism where tens of thousands of people visit the same place at the same time of year 35
mean the sum of the quantities divided by the number of quantities 7
meander winding curve or bend in a river 58–59
meander belt area of a drainage basin where meanders are formed 56–57
medial (moraine) merging of two lateral moraines 45
median the point in a series that is exactly in the middle of the values 7
mega-city a city with over 10 million inhabitants 68
meltwater water resulting from melting snow and ice 45
mid-latitudes the temperate zones between the tropics and polar regions 18
millionaire city a city with over a million inhabitants 68
mitigation measures to reduce the causes of climate change, e.g. cutting greenhouse gas emissions; eliminating long-term risk to human life and property; international targets to reduce greenhouse gas emissions; carbon capture 21
modal class the class with the highest frequency 7
mode the most frequently occurring data value in a series 7
moraine material carried and deposited by ice 45
mouth where a river ends; usually meeting the sea 56, 59
mudflats low-lying area of silt, clay and sand particles found in low energy environments, exposed at low tide 59
NEE newly emerging economy 68, 70, 76, 80
negative multiplier effect where a negative event triggers other negative events 83
North Atlantic Drift a powerful warm ocean current 18
northing line one of the horizontal lines crossing an OS map from one side to the other; they are called northings as the numbers increase to the north 4
notch small indentation in a cliff face at high tide mark; results from wave action 52
nucleated a pattern or distribution is nucleated if there is a concentration of something in a tightly-confined cluster, e.g. a nucleated settlement pattern 4
nutrient cycle how the minerals that provide energy and sustenance to living organisms circulate around an ecosystem; minerals are moved between the atmosphere, biomass, litter layer and soil 27
offshore away from the land 53, 54–55
organic produced without the use of artificial chemicals 92, 97
orientation direction, using compass points or bearings 5
out-migration moving away from a local community 33
overland flow water moving over the surface towards the channel 60–61
ox-bow lake formed when a river meander is cut off 59
pastoral farming farming aimed at producing livestock rather than crops 35, 48
percentile each of the 100 equal groups into which a population can be divided according to the distribution of values of a particular variable 7
peat partly decomposed vegetation and organic matter; holds a lot of water 61
peripheral describes places which are distant from the centre or core of activity 4
permaculture the development of agricultural ecosystems intended to be sustainable and self-sufficient 97
permafrost permanently frozen layer below the Earth's surface 36
permeable a rock or surface that allows water to pass through 6
physical water scarcity a situation where the water is not abundant enough to meet all demands for the population 98
plate margin the boundary of the plates that form the upper layer of the Earth 8
plateau flat, elevated land rising sharply above the area surrounding it on at least one side 5, 58

plucking erosion process; weakened rock removed from bedrock by a glacier **45, 46**

plunging destructive waves that bring material from a beach towards the sea **50**

political map shows boundaries linked to areas of governance such as countries, states, and counties, and the location of major cities **4**

pollution the introduction by humans of harmful material into the environment; always specify what type of pollution: air pollution; water pollution, etc. **49**

prevailing wind a wind from the direction that is most usual at a particular place or season **18**

primary data data that you and your fellow students have collected **109**

primary effects the effects that occur immediately after the natural disaster happens, e.g. the ground shakes **10–11, 12–13**

problem-solving thinking about the possible solutions to a problem and deciding which course of action to take **110**

producer an organism, e.g. a tree, that makes its own food via photosynthesis **26–27**

proxy measures indirect ways of measuring variables, such as tree ring growth as an indicator of temperature in past years **20**

pull factors positive factors attracting people to live in a place **69**

push factors negative factors making people want to leave a place **69**

pyramidal peak sharp-pointed, frost-shattered mountain top with corries on at least three sides **46**

pyroclastic flow a high temperature avalanche of gas, ash, cinder and rock that rushes down the slopes of a volcano at speeds of up to 450 mph **12–13**

qualitative data non-numerical data which might involve subjective judgements, e.g. field sketches; photographs; video; quotes or opinion **109**

quality of life the wide range of human needs that should be met alongside income growth such as access to housing, education and health, nutrition, security, happiness, etc. **76**

quantitative data numerical data or statistics, e.g. width and depth of a river channel; an environmental quality survey **109**

quartiles each of four equal groups into which a population can be divided according to the distribution of values of a particular variable **7**

Quaternary period the last 2.6 million years in which there have been several glacials **20**

radial pattern spread out in a linear way from a central point, like the spokes on a bicycle wheel or rays of the sun, e.g. rivers or glaciers flowing in all directions from high ground; transport links from a hub **4, 6**

rain shadow a dry area on the leeward side of a hilly/mountainous area **18**

regeneration reviving old run-down urban areas by either improving what is there or clearing it and rebuilding **73**

relief the difference in elevation in an area; shape of the land **4, 6, 48**

relief map shows the height and shape of the land (topography) using symbols such as contours or layer shading **4**

relief rainfall occurs when moist air rises over a physical barrier, e.g. mountains **18**

renewables a natural resource and source of energy that is not depleted by use, such as water, wind or solar power **93**

reprofiling change the face/front of a slope; could be to make more or less steep, higher or lower, than originally **55**

resilience capacity to cope with a hazard such as a tropical storm, often dependent on how effectively the 3 Ps are implemented **17**

resistant tough; able to withstand weathering and/or erosion **50**

ribbon lake long narrow lake formed by a glacier **46–47, 48–49**

Richter scale a measure of the energy released by an earthquake **10–11**

ridge mountains forming a continuous elevated crest **5**

river bank the land alongside a river, above the channel **57**

salinisation the increase in the salt content of water **31, 35**

saltation bouncing or jumping movement of particles along a surface **51**

saltmarsh area of deposition of fine material that is tidal; adapted plants grow and stabilise the area **55**

sampling techniques data is collected from a small part of the population or a small number of sites and used to inform what the whole picture is like. There are three main types of sampling: random; systematic; stratified **109**

sand dune accumulation of sand in hill form at the back of beaches along low-lying coastlines **53**

saturated no more water can be held; if a permeable material becomes saturated, it becomes impermeable **60**

scale ratio which compares a measurement on a map between places to the actual distance between these places in reality; for example, the most commonly used OS maps use a scale of 1:50000 and 1:25000; scale is also used when referring to places or areas of different sizes, e.g. local, regional, national, international, global scale **5**

sea couch salt-loving plant; one of the first to grow on sand dunes **55**

secondary data data collected by another person, group or organisation which you have accessed via articles, websites, books, etc. **109**

secondary effects the effects that occur subsequent to the primary effects, such as gas mains rupturing **10–11, 12**

settlement a place where people establish a community **4, 17, 32, 34, 37, 70–71, 72, 75**

shield volcano large sized, low-lying volcano formed from fluid magma flows **12**

shifting cultivation ground is cultivated and then abandoned to allow its fertility to be restored **32**

short-term aid immediate relief after an emergency **79**

silage crops set aside for farm animals in winter **48**

sinuous curving movements in a river's flow **59**

slip-off slope area of deposition at the inside bend of a meander **59**

soft river engineering range of measures used to manage river flow and floods which do not use materials such as concrete, stone or steel, e.g. river restoration, floodplain zoning; they allow natural processes to occur and don't alter the natural environment; often less expensive than hard engineering and considered to be more economically, socially and environmentally sustainable **6, 61**

solution dissolving action in water **50, 57**

source beginning of a river **56**

sparse a pattern or distribution is sparse if it shows a low density and/or is very dispersed in an area, e.g. sparse drainage pattern; sparse population **4**

spilling wave which pushes material onshore **50**

spine the main channel in a river system; also known as the trunk **56**

spit long protrusion of deposited material extending from the coast into the sea **53**

spot heights points of known height shown using a dot with a number beside it, usually in metres above sea level; they appear on a map, but not on the landscape **5**

spur a piece of land jutting into a river or stream **5**

stakeholders people or groups with a real interest in an issue or in the development of a decision or policy **110–111**

steep having a noticeable slope of over 30° **60**

storm surge a rising of the sea as a result of wind and atmospheric pressure changes associated with a storm **15, 17**

subdued description of a river which responds very slowly to rainstorms **60**

sub-glacial (moraine) debris under the ice **45**

subsistence farming crops produced solely for the farmer and family **32**

supra-glacial (moraine) debris on top of the ice **45**

surging wave which pushes material up a beach, making it steeper **50**

suspended load particles carried along within the body of water (river or sea) **57**

sustainable meeting the needs of the present population without compromising the needs of future generations **33, 74, 85, 97, 100, 106**

swash forward wave motion **50**

symmetrical similar gradient on both sides **6**

synoptic thinking the application of knowledge, understanding and skills from all of your geographical learning to a new situation **110**

terminal (moraine) crescent-shaped deposit at the furthest reach of a glacier **45, 47**

terraced fields used for farming in hilly regions **35**

terracing making a slope into shallow steps; this makes the overall slope less steep and makes flat surfaces to hold water **61**

terrain landscape pattern **48**

thalweg the fastest part of a river, where there is least friction **58**

throughflow lateral movement of water below the surface (through the soil towards the river) **60**

tidal influenced by the sea; seawater advances and retreats with the tide **59**

till clay-containing rock fragments deposited from under a glacier **45, 47**

tourism activities of visitors to an area and those who cater for them **34–35, 49, 80–81**

transnational corporation (TNC) enterprise operating in at least two countries **80, 102**

traction rolling and sliding of larger, heavier particles along a river bed or in the sea **51, 57**

transpiration water evaporates from plants into the atmosphere **31, 60**

transportation material being carried away; in rivers, by solution, suspension, traction and saltation **45, 57**

triangulation pillars points of very precise known height shown using a dot with a number beside it, usually in metres above sea level; these appear on a map as a triangular symbol, and are marked on the landscape with a concrete pillar **5**

tributary any river flowing into another, bigger river **46, 56**

tropical storm low pressure systems with distinct structure and features formed over warm ocean waters in low latitudes **14–15, 16–17**

trough end upstream end of a glaciated valley with a sudden closure **46**

truncated spur former interlocking spurs from a river valley which have been cut away by glacial erosion **46, 49**

trunk the main channel in a river system; also known as the spine **56**

tsunami a large wave caused by an earthquake, volcanic eruption or coastal landslide **10–11**

tundra vast treeless zone in the arctic regions of the world **29, 36–37**

typhoon tropical storms in the western North Pacific, e.g. Philippines, Japan **14–15, 16–17**

undermine erode from below; cause weakness **57**

urbanisation an increase in the number of people living in towns and cities compared with rurally **68–69**

U-shaped valley deep, steep-sided valley formed by glaciers; also known as a trough **46**

valley between areas of higher land, often with a river running through **5, 6, 8, 46–47**

waterborne diseases caused by coming into contact with an infected water source **99, 101**

water conservation beneficial reduction of water usage **100**

water deficit not enough rainfall, leaving a water shortage **19**

waterfall where river water flows over a sudden vertical break **46, 58**

waterlogged saturation of the soil by an excess of water **48, 61**

watershed the dividing line between two drainage basins **56**

water security the reliable availability of water to sustain the population **98**

water stress lacking water **19**

water surplus having more water than needed **19**

water table level below ground at which rocks are saturated **53, 60**

water transfer scheme moving water from an abundant river basin to one where water is less available **100**

wave cut platform gently shelving area of solid rock stretching out to sea from the cliff front **52**

weathering break-down of rock as a result of exposure to the atmosphere **44–45, 46–47, 50, 52**

wetted perimeter length of channel bed in contact with water **56, 60**

wilderness areas untouched by human influence **37**

wind patterns the semi-predictable movement of masses of air around the world, e.g. the trade winds **28**

yield the measure of crops produced, often in kgs per hectare **95, 96**

Collins

AQA GCSE 9-1 Revision

Geography

Geography

AQA
GCSE 9-1

Workbook

Janet Hutson, Dan Major, Paul Berry,
Brendan Conway and Robert Morris

Revision Tips

Rethink Revision

Have you ever taken part in a quiz and thought *'I know this!'* but, despite frantically racking your brain, you just couldn't come up with the answer?

It's very frustrating when this happens but, in a fun situation, it doesn't really matter. However, in your GCSE exams, it will be essential that you can recall the relevant information quickly when you need to.

Most students think that revision is about making sure you ***know** stuff*. Of course, this is important, but it is also about becoming confident that you can ***retain*** that *stuff* over time and ***recall*** it quickly when needed.

Revision That Really Works

Experts have discovered that there are two techniques that help with all of these things and consistently produce better results in exams compared to other revision techniques.

Applying these techniques to your GCSE revision will ensure you get better results in your exams and will have all the relevant knowledge at your fingertips when you start studying for further qualifications, like AS and A Levels, or begin work.

It really isn't rocket science either – you simply need to:

- **test yourself** on each topic as many times as possible
- **leave a gap** between the test sessions.

Three Essential Revision Tips

1. **Use Your Time Wisely**
   - Allow yourself plenty of time.
   - Try to start revising at least six months before your exams – it's more effective and less stressful.
   - Your revision time is precious so use it wisely – using the techniques described on this page will ensure you revise effectively and efficiently and get the best results.
   - Don't waste time re-reading the same information over and over again – it's time-consuming and not effective!
2. **Make a Plan**
   - Identify all the topics you need to revise (this All-in-One Revision & Practice book will help you).
   - Plan at least five sessions for each topic.
   - One hour should be ample time to test yourself on the key ideas for a topic.
   - Spread out the practice sessions for each topic – the optimum time to leave between each session is about one month but, if this isn't possible, just make the gaps as big as realistically possible.
3. **Test Yourself**
   - Methods for testing yourself include: quizzes, practice questions, flashcards, past papers, explaining a topic to someone else, etc.
   - This All-in-One Revision & Practice book provides seven practice opportunities per topic.
   - Don't worry if you get an answer wrong – provided you check what the correct answer is, you are more likely to get the same or similar questions right in future!

Visit our website to download your free flashcards, for more information about the benefits of these techniques, and for further guidance on how to plan ahead and make them work for you.

www.collins.co.uk/collinsGCSErevision

Contents

The Challenge of Natural Hazards

Refer to this map for questions 1–3.

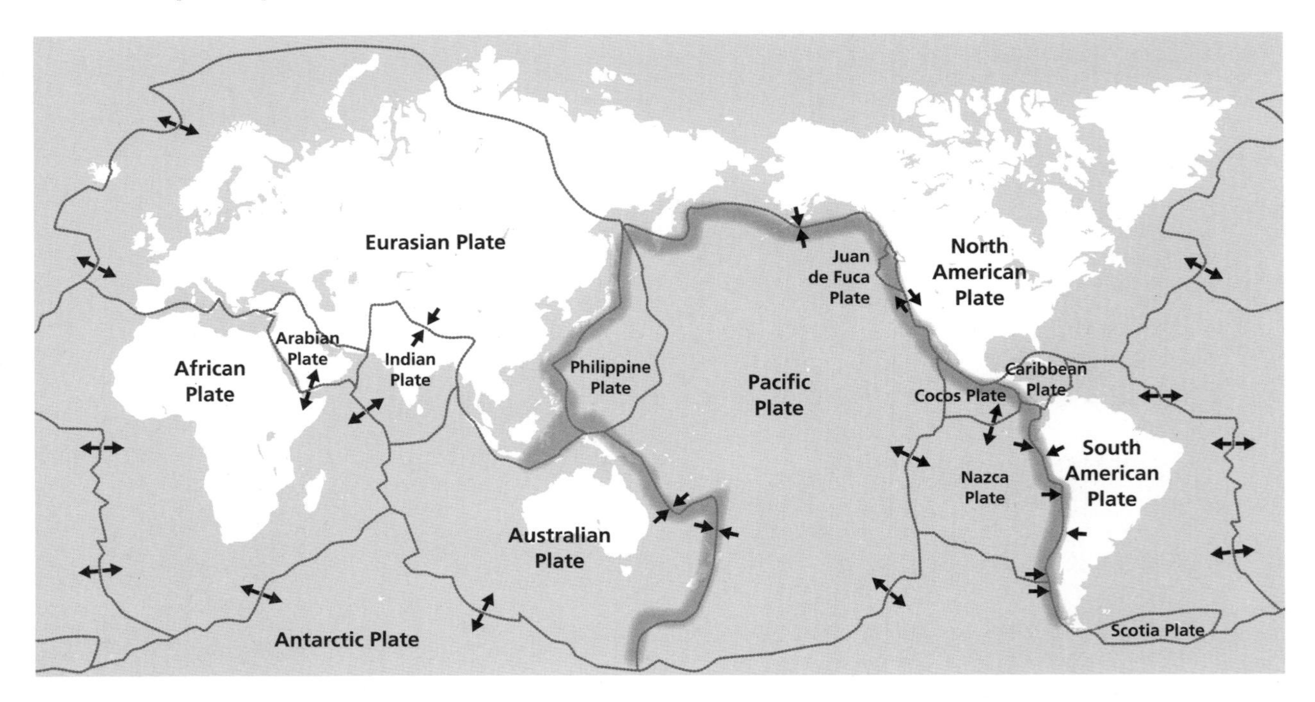

1. What is the common name given to the highly active area of tectonic activity highlighted on the map? [1]

2. Use the map to help you name two plates that form a constructive margin. [2]

3. Use the map to help you name two plates that form a destructive margin. [2]

4. Name an example of a major earthquake and state where and when it happened. [2]

5. What is the difference between the focus and the epicentre of an earthquake? [2]

6. Describe ways of predicting when and where an earthquake might happen. [6]

7. Describe the contents of an earthquake survival kit for a typical householder.

   Explain how three items could help in an emergency. [6]

8 With reference to the Haiti 2010 earthquake, state whether the following statements are true or false: [5]

a) A nuclear power station was severely damaged. ....

b) The epicentre was within 15 miles of the capital city. ....

c) It measured over 6 on the Richter scale. ....

d) It produced tsunami waves up to 33 feet in height. ....

e) It occurred at night while people were asleep. ....

9 Describe the effects of a tsunami. [6]

10 Describe some of the problems faced after an earthquake. [6]

11 Name an example of a major volcanic eruption and state where and when it happened. [2]

12 With reference to the Merapi volcanic eruption of 2010, state whether the following statements are true or false: [5]

a) It occurred at a destructive plate margin. ....

b) It produced a giant ash plume. ....

c) It resulted in a major pyroclastic flow. ....

d) It produced a serious lahar. ....

e) It occurred beneath an ice cap. ....

13 Which of the following is the main variable used by the Saffir–Simpson scale to measure the intensity of a tropical storm? Tick the correct answer. [1]

A Duration of the storm in hours ☐

B Width as measured from space ☐

C Rainfall amount ☐

D Wind speed ☐

14 **a)** Describe the pattern of tropical storm tracks in the map below. [4]

Tropical Storm Tracks Over the Last 70 Years

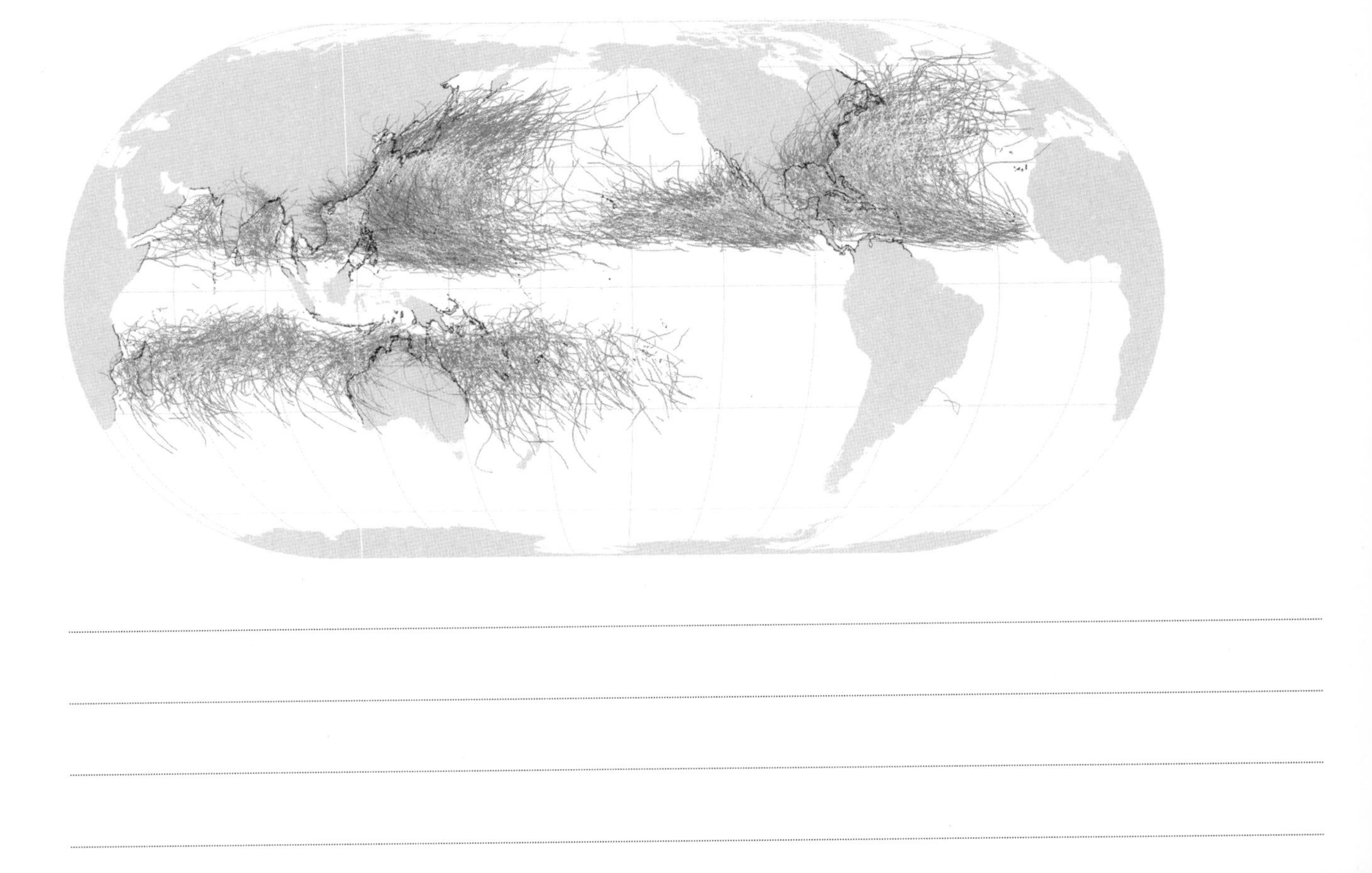

....

....

....

....

b) Suggest reasons for the links between the distribution of tropical storms and sea surface temperatures. [4]

Average Sea Surface Temperatures

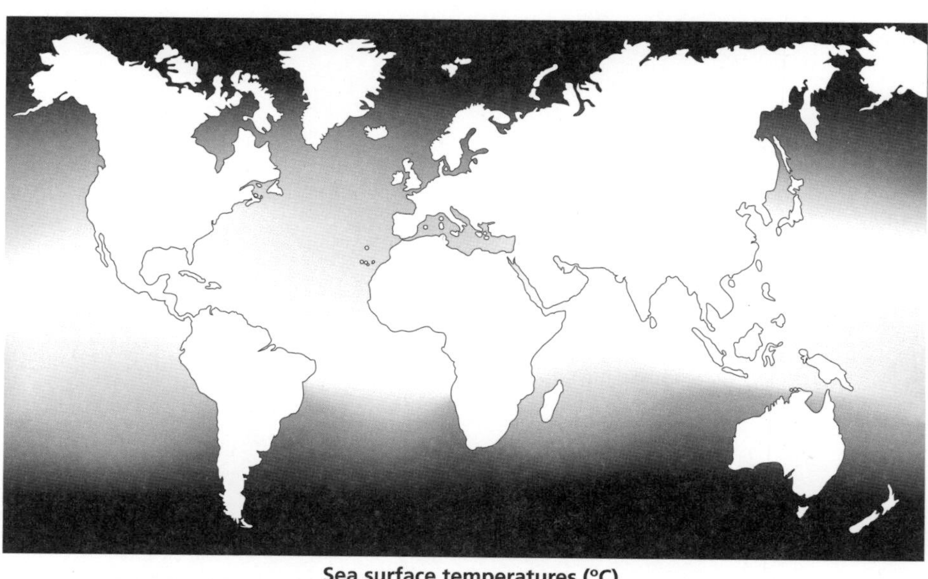

....

....

....

....

15 How did the long-term response to Typhoon Haiyan in 2013 improve the response to Typhoon Hagupit in 2014? [2]

....

....

16 Summarise the maritime influences on the UK climate. [2]

....

....

17 In what ways can the UK climate be influenced by continental air? [2]

18 The chart shows changes in carbon dioxide from the Mauna Loa Observatory in Hawaii.

Why is this location considered to be a useful place to measure average carbon dioxide in the atmosphere? [2]

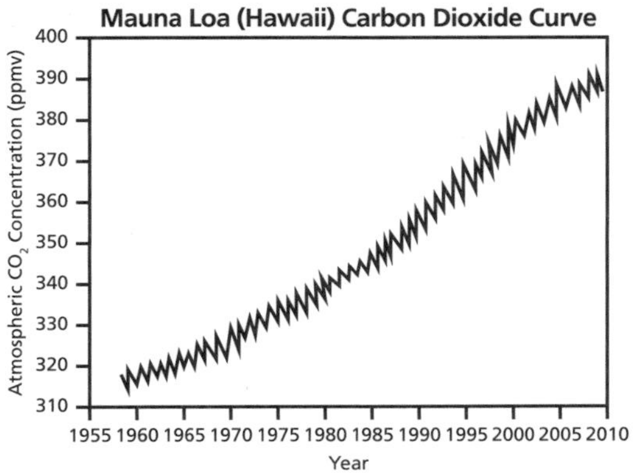

19 Explain some key differences between the causes of climate change before 1850 and since 1850. [8]

Total Marks .... / 70

The Living World

1. Describe one example of how human action can disrupt the balance within a particular ecosystem. [3]

2. Give two places where high pressure cells can be found in July. [2]

3. Give an example of a place where the saguaro cactus grows naturally. [1]

4. Define 'ecotourism'. [2]

5. How has mass tourism increased water use in southern Spain? [2]

6. What recent development in tourism has been brought about by the desire among tourists to experience new, undiscovered and potentially hazardous environments? [1]

7. Buttress roots give stability for rainforest trees but also fulfil what other critical function? [2]

8. Define 'irrigation'. [1]

9 Using named examples, describe the opportunities for development found in hot deserts. [6]

10 Define 'desertification'. [1]

11 How can the process of desertification be tackled? [4]

12 Which of the following places could be described as being semi-arid? Tick the correct answer. [1]

A The Sahel region of central Africa ☐

B The north Italian plain ☐

C The Allegheny plateau of the USA ☐

D The Thames river basin ☐

13 Define 'appropriate technology'. [2]

Total Marks .... / 28

Physical Landscapes in the UK

1. How would you know that an area of rock had undergone freeze-thaw action? [2]

2. If friction causes ice to melt, how can abrasion take place beneath a moving glacier? [2]

3. Explain the role of water in plucking. [3]

4. How does the bedrock type and hardness affect glacial erosion? [3]

5. What shape is a terminal moraine? [1]

6. What is the relationship of an arête to a corrie? [2]

7. Why is pastoral farming more successful than arable in upland areas? [2]

8. Why are mining and quarrying less important today than in the past in places like the Lake District or Snowdonia? [3]

9. Describe diversification on a farm in a glaciated upland area. [3]

10. How do recreation activities negatively affect the landscape? [3]

11. How will soils on a U-shaped valley floor differ from those on the valley sides? [4]

12 Look at the photograph below showing part of a coast that is being managed.

a) Are the coastal management measures shown, hard or soft engineering? [1]

b) Give the names of the structures shown. [2]

c) Explain how these structures are protecting this section of coast. [4]

13 What type of mass movement is most likely on clay cliffs? [1]

14 Give two benefits to humans of managed retreat. [2]

15 Describe an effect of weathered material from a cliff remaining at the cliff foot. [2]

....

....

16 Look at the photograph below, which shows waves approaching a vegetated area on the coast of the Isle of Mull, Scotland.

a) What effect might the vegetation have on the rocks? [2]

....

....

b) Describe **one** type of wave erosion. [2]

....

....

17 a) For a named section of coastline in the UK that you have studied, explain why coastal protection was considered to be necessary. [3]

b) Describe one of the methods of coastal protection used on that same section of coastline. [2]

18 Why would deposition of bedload reduce a river's power to erode? [2]

19 Describe and explain the change in a river channel that could increase the possibility of erosion taking place. [4]

20 Explain the formation of floodplains. [4]

21 Describe the formation of a waterfall resulting from faulting of rocks. [4]

....

....

....

....

22 For a UK river you have studied, choose one named landform caused by erosion, and describe its location and key features. [4]

....

....

....

....

23 Describe the formation of an ox-bow lake. [4]

....

....

....

....

24 Many places have embankments as part of their flood defence. Describe **one** positive feature and **one** negative feature of these embankments. [4]

....

....

....

....

Total Marks .... / 75

Urban Issues and Challenges

1. What is a 'millionaire city'? [1]

2. What is a 'mega-city'? [1]

3. As urban areas grow, they encroach into rural (countryside) areas.

   What term is used to describe this process? [1]

4. Push factors and pull factors can cause urban areas to grow. Identify the following as either push factors or pull factors: [4]

   a) Higher wages

   b) Unemployment

   c) Isolation

   d) Better schools and hospitals

5. How did local residents benefit from the London Olympics in 2012? [4]

6 What are the distinctive features of sustainable urban living? [6]

7 Evaluate strategies used to make urban transport more sustainable. [5]

Total Marks .... / 22

The Changing Economic World

1. Identify three reasons why world death rates are falling. [3]

2. Identify three reasons why world birth rates are falling. [3]

3. Suggest two reasons why birth rates remain higher in lower income countries (LICs). [2]

4. Which of the following statements are true for a higher income country (HIC)?
   Tick the correct answers. [2]
   - A Low birth rates ☐
   - B High death rates ☐
   - C High elderly population ☐
   - D High infant mortality ☐
   - E Rapidly increasing population ☐

5. Name two jobs that make up the tertiary sector. [2]

6. Name two jobs that make up the quaternary sector. [2]

7. Describe an example of how primary industry can have a negative effect on the environment. [2]

8 Describe an example of how secondary industry can have a negative effect on the environment. [2]

....

....

9 Describe an example of how tertiary industry can have a negative effect on the environment. [2]

....

....

10 What are the advantages of transnational corporations (TNCs) locating in a newly emerging economy (NEE) like Vietnam:

a) for the TNC? **b)** for the host country? [5]

....

....

....

....

....

11 What is 'intermediate technology'? Describe an example in detail to explain how it can improve people's lives. [6]

....

....

....

....

....

....

....

Total Marks .... / 31

The Challenge of Resource Management

1. What is a 'fossil fuel'? [1]

2. Suggest reasons why the increased use of nuclear energy in the UK is likely to cause controversy. [3]

3. Why is increasing food security in LICs particularly hard to solve? [4]

4. What is meant by 'food miles' and 'carbon footprint'? [4]

5. Why do more than one billion people lack access to safe drinking water? [3]

6 Study the figure below. Suggest reasons why some regions of the world have water scarcity. [4]

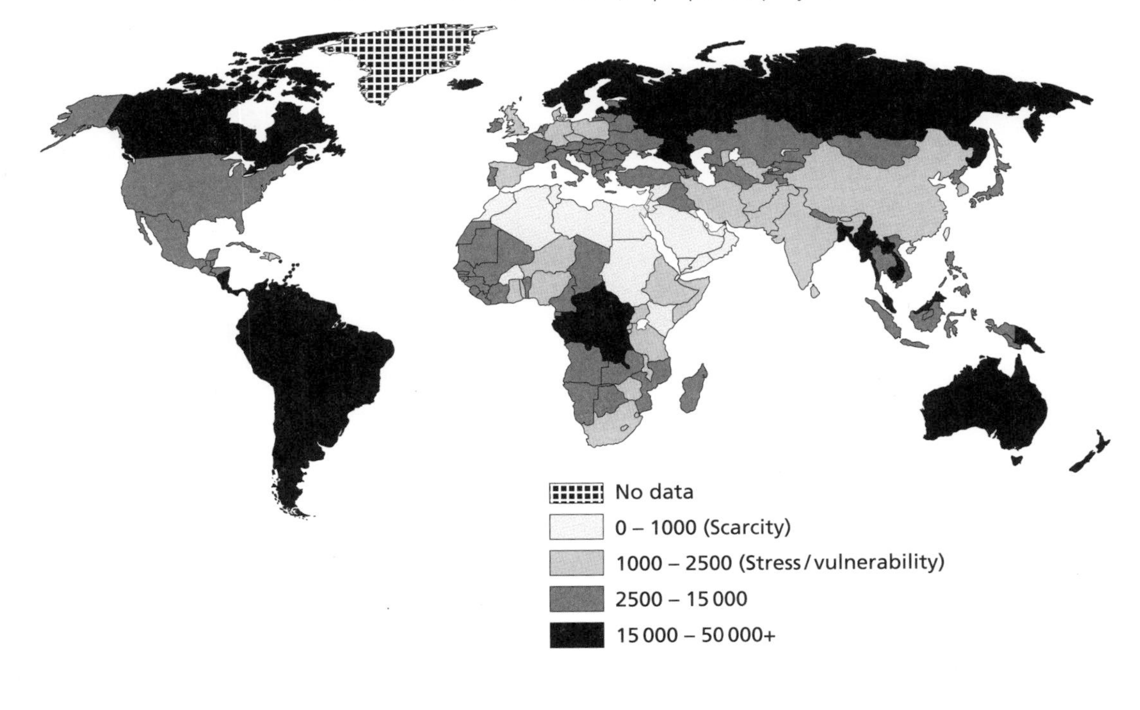

7 Define 'energy mix'. [1]

8 Explain how LICs can improve their energy security. [3]

9 Explain why the rising price of oil leads to higher food prices. [5]

10 How can the design of buildings lead to savings in energy costs? [2]

11 Define the term 'sustainable transport system'. [3]

12 Why is biomass a sustainable energy source? Give examples. [3]

13 How has technology been used to reduce fuel costs for transport and energy production? [4]

14 'Adaptation is an effective way of reducing usage of energy.'

What is meant by adaptation? [1]

15 What is meant by 'greywater harvesting'? Give examples. [3]

16 Give some advantages and disadvantages of exploiting groundwater resources. [4]

Total Marks .... / 48

Geographical Applications

1 For your geographical enquiry, describe and explain patterns in your data and any anomalies which did not correspond to the main patterns. [9 + 3 SPaG]

Total Marks .... / 12

Collins

GCSE
GEOGRAPHY

Paper 1 Living with the physical environment

Time allowed: 1 hour 30 minutes

Materials

For this paper you must have:

- a pencil
- a ruler.

Instructions

- Use black ink or a black ball-point pen.
- Answer **all** questions in Section A and Section B.
- Answer **two** questions in Section C.
- Cross through any work you do not want to be marked.

Information

- The marks for questions are shown in brackets.
- The total number of marks available for this paper is 88.
- Spelling, punctuation, grammar and specialist terminology will be assessed in Question 01.6.

Advice

- For the multiple-choice questions, completely fill in the circle alongside the appropriate answer(s).

  CORRECT METHOD  WRONG METHODS  

- If you want to change your answer, you must cross out your original answer as shown. 
- If you wish to return to an answer previously crossed out, ring the answer you now wish to select as shown.

Name: ....

Practice Exam Paper 1

Section A The challenge of natural hazards

Answer **all** questions in this section

Question 1 The challenge of natural hazards

Study **Figure 1**, a map showing the major tectonic plates and their direction of movement.

Figure 1

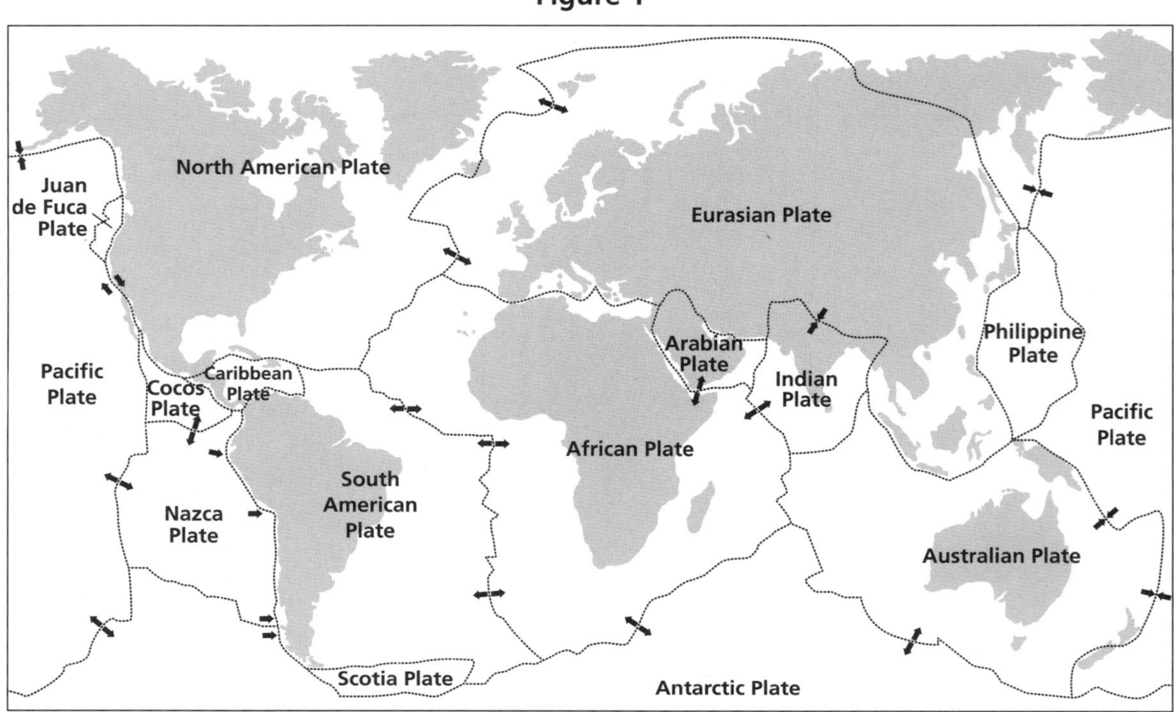

0 1 · 1 Which **two** words could be used to complete the following statement? Shade **two** circles only.

The plate boundary shown between the Nazca Plate and Pacific Plate is: **[2 marks]**

A convergent ○

B divergent ○

C constructive ○

D destructive ○

E collision ○

F conservative ○

0 1 · 2 Complete the following sentence.

An example of a conservative plate boundary is the boundary between the

.... plate and the .... plate. **[1 mark]**

Study **Figure 2**, a photograph taken in Nepal following a magnitude 7.8 earthquake in April, 2015.

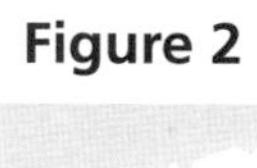

Figure 2

0 1 · 3 Explain what is meant by the term 'magnitude 7.8'. **[2 marks]**

....

....

0 1 · 4 Identify the impact on the landscape shown in **Figure 2**. **[1 mark]**

....

0 1 · 5 Suggest **two** ways in which the impact shown in **Figure 2** could affect people living in the area. **[4 marks]**

....

....

....

....

Question 1 continues on the next page

Study **Figure 3a**, an extract about Mount Etna, a volcano on the Italian island of Sicily, and **Figure 3b**, a photograph showing a house in Sicily following a recent eruption.

Figure 3a

Etna produces pyroclastic flows, ash falls and mudflows. However, the most hazardous type of activity is lava flows, in particular to Catania, a city of 300 000 people and the second largest in Sicily. Lava flows do not usually move fast enough to present danger to the population, but they can cover large areas and destroy crops and buildings. The area around Etna could be evacuated only with great difficulty.

Figure 3b

0 1 · 6 Describe and explain the possible problems and benefits of living near an active volcano. Use **Figure 3a** and **3b** and an example you have studied.

[9 marks] [+ 3 SPaG marks]

Study **Figure 4a**, a photograph showing Tadcaster, North Yorkshire, in December 2015, and **Figure 4b**, a statement issued by the Environment Agency shortly afterwards.

Figure 4a

Figure 4b

Widespread heavy showers bring a risk of flooding from rivers and surface water across large parts of the country. December 2015 was the wettest calendar month in the UK since 1910, meaning many of these showers are falling on saturated ground and full rivers. We urge people to remain vigilant and check flood alerts and warnings online.

0 1 · 7 Using **Figures 4a** and **4b**, identify **two** pieces of evidence that the rainfall experienced in winter 2015–2016 was unusual. **[2 marks]**

....

....

Question 1 continues on the next page

Practice Exam Paper 1

Study **Figure 5**, a diagram showing a cross-section of a hurricane (tropical storm).

Figure 5

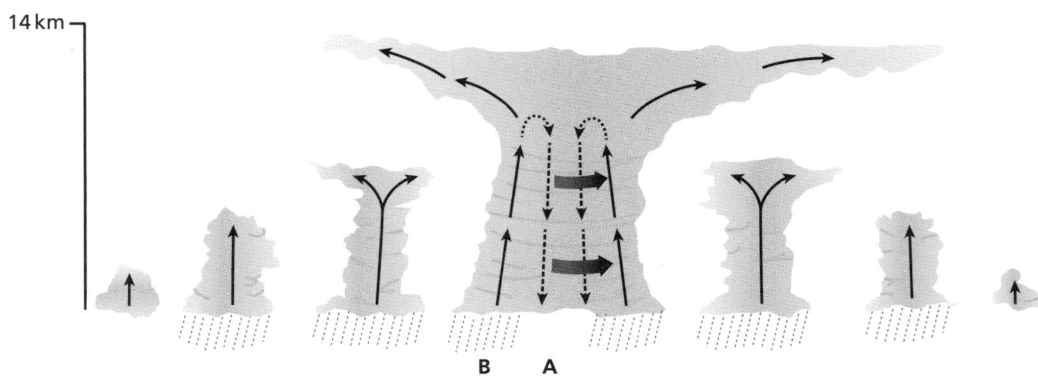

0 1 · 8 State the feature of a hurricane labelled **A** in **Figure 5**. **[1 mark]**

....

0 1 · 9 Contrast the weather conditions that would be experienced on the ground at Area A compared to Area B. **[2 marks]**

....

....

There are a number of types of evidence which suggest that the Earth is experiencing climate change, including:

glaciers and ice sheets **sea level** **environment**

0 1 · 1 0 Choose **one** of the three evidence types shown above.

State your choice of evidence and explain how it might show that the Earth is experiencing climate change. **[2 marks]**

....

....

0 1 · 1 1 Describe how climate change could have a **negative impact** on economic activities. **[4 marks]**

....

....

....

....

Section B The living world

Answer **all** questions in this section

Question 2 The living world

Study **Figure 6**, a world map showing some large-scale ecosystems (biomes).

Figure 6

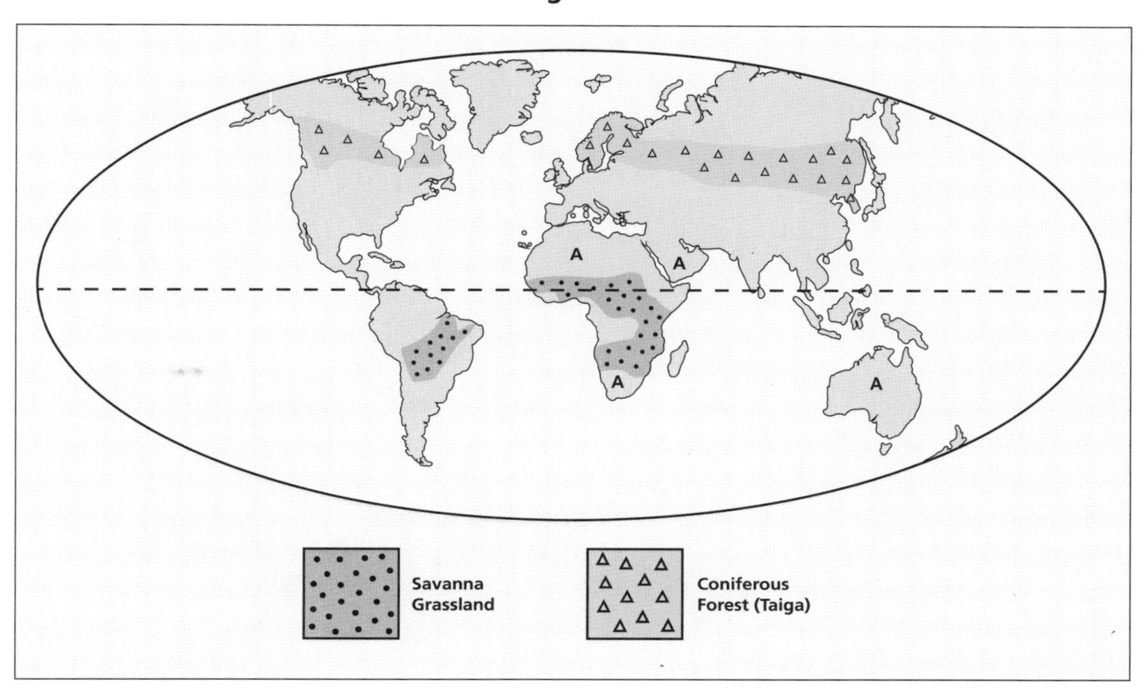

0 2 · 1 Using **Figure 6** and **your own knowledge**, choose the correct words to complete the following sentences.

Circle the correct answer from those given in bold.

Coniferous forests are found **mainly/entirely** in the Northern Hemisphere.

They are located **north/south** of tundra regions.

There is **some/no** savanna grassland in Asia. **[3 marks]**

0 2 · 2 State which type of large-scale ecosystem you would expect to find in the areas labelled **A** on **Figure 6**. **[1 mark]**

....

Question 2 continues on the next page

Study **Figure 7**, a diagram showing a simple food chain.

Figure 7

0 2 · 3 Add arrows to **Figure 7** to show the direction of energy flow. **[1 mark]**

0 2 · 4 Which of the four organisms shown in **Figure 7** is classed as a producer? **[1 mark]**

....

Study **Figure 8**, a diagram showing the structure of a rainforest, with some of the strata labelled.

Figure 8

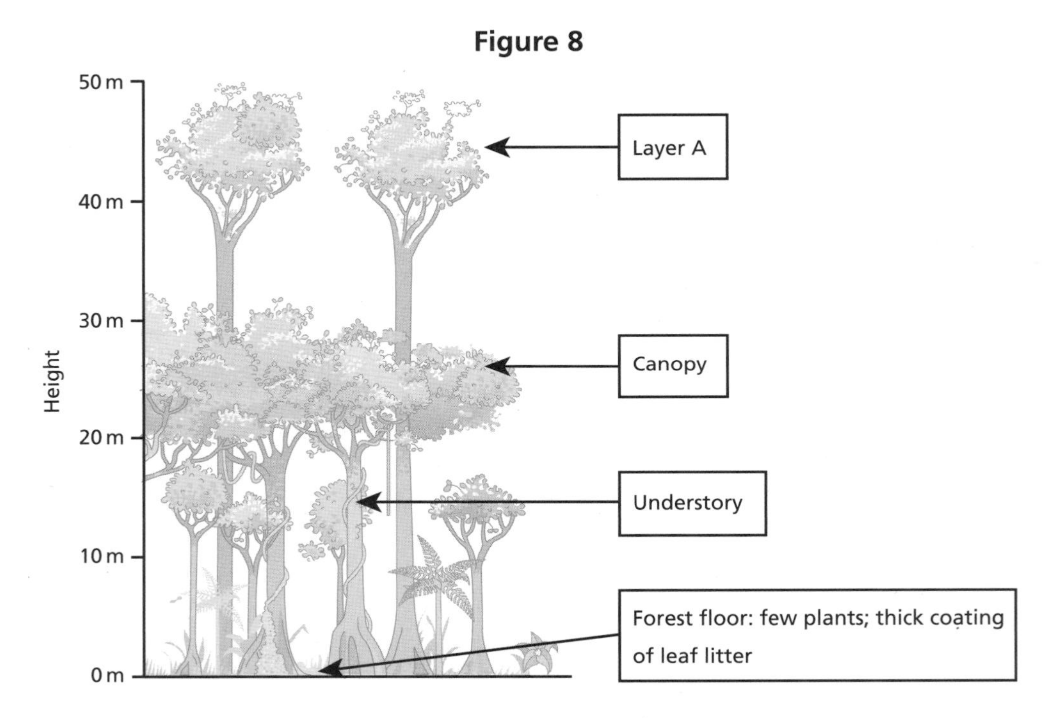

0 2 · 5 Identify the layer labelled **A** on **Figure 8**. **[1 mark]**

....

0 2 · 6 Explain why the forest floor has lots of leaf litter but few plants. **[2 marks]**

....

....

Study **Figure 9**, a diagram of a leaf from a tree in a tropical rainforest.

Figure 9

0 2 · 7 Choose **one** of the characteristics labelled on **Figure 9**.

State the characteristic you have chosen and explain how it enables the plant to adapt to the rainforest climate and weather conditions. **[1 mark]**

....

....

0 2 · 8 Describe the impact of removing trees from the rainforest on the weather, soil and rivers. **[6 marks]**

....

....

....

....

....

....

0 2 · 9 For a hot desert environment or cold environment you have studied, discuss the challenges presented to humans living in the region and explain how they have been overcome. **[9 marks]**

....

....

....

....

....

....

....

....

....

End of Section B **Turn over for Section C**

Practice Exam Paper 1

Section C Physical landscapes in the UK

Answer **two** questions from the following:

Question 3 (Coasts), Question 4 (Rivers), Question 5 (Glacial)

Shade the circle below to indicate which two optional questions you will answer.

Question	0	3	⬭

Question	0	4	⬭

Question	0	5	⬭

CORRECT METHOD ⬬ WRONG METHODS

Question 3 Coastal landscapes in the UK

Study **Figure 10**, a photograph from a beach in Suffolk.

Figure 10

0 3 · 1 Describe the erosion process that produces the rounded beach material shown in **Figure 10**. **[2 marks]**

....

....

Study **Figure 11**, a photograph of Dunwich in Suffolk.

Figure 11

0 3 · 2 Identify the process of mass movement that has taken place at **A** on **Figure 11**. **[1 mark]**

....

0 3 · 3 State a rock type that is often associated with the process illustrated by **Figure 11**. **[1 mark]**

....

Study **Figure 12**, a photograph showing a section of the coastline of the Isle of Anglesey.

Figure 12

0 3 · 4 Identify the feature that is labelled **B** on **Figure 12**. **[1 mark]**

....

Question 3 continues on the next page

Practice Exam Paper 1

Study **Figure 13**, a 1 : 25 000 map extract of part of the coastline of the Isle of Mull.

Figure 13

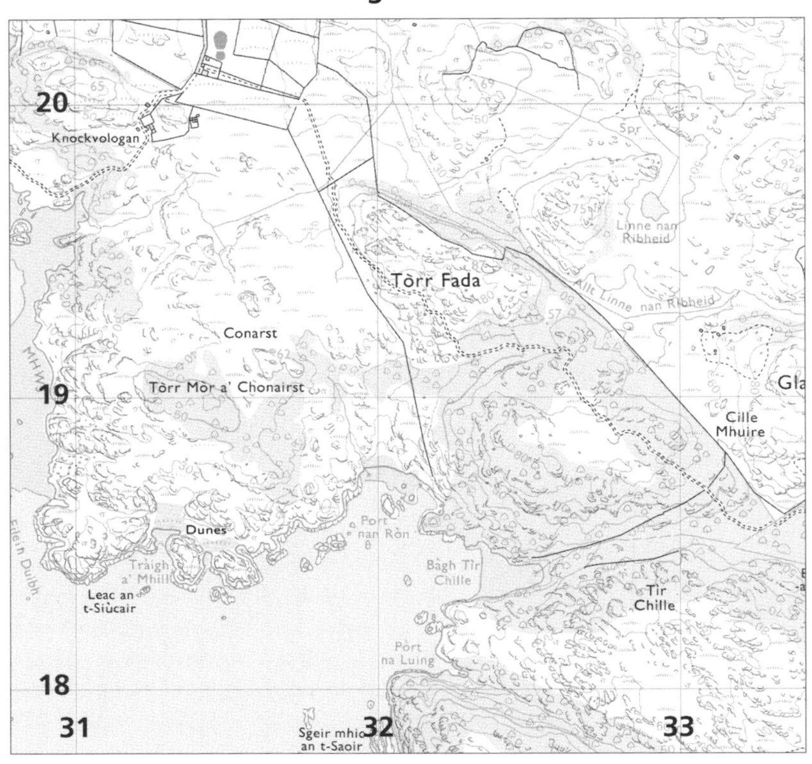

0 3 · 5 Using **Figure 13**, match each coastal feature below to the correct grid reference.

Write the letter of the correct grid reference in the table.

Choose from the following grid references:

A 318184 **B** 321195 **C** 312182 **D** 324184

Feature	Grid reference (A, B, C or D)
Stack	
Sandy beach	

[2 marks]

0 3 · 6 State **two** characteristics of a constructive wave. **[2 marks]**

Study **Figure 14**, a diagram showing a spit, a coastal deposition feature that results partly from longshore drift.

Figure 14

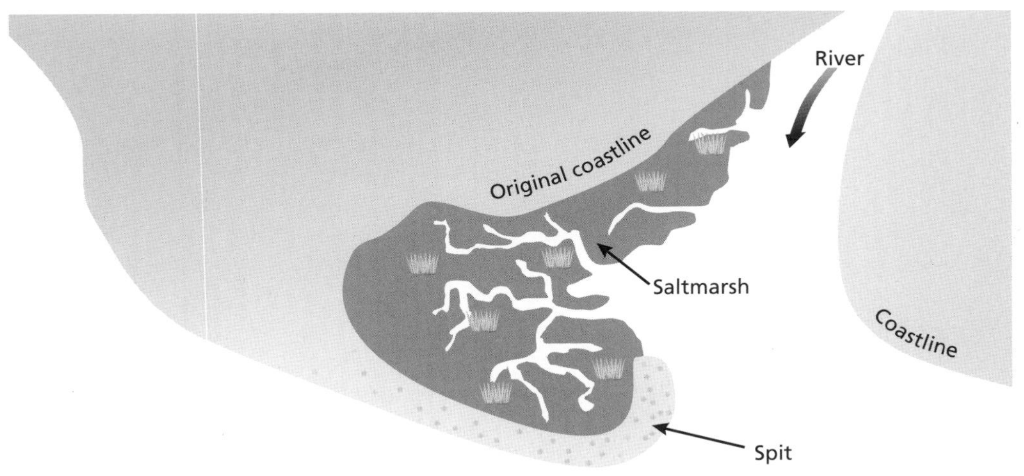

0 3 · 7 Give a named and located example of a spit and describe its formation. **[6 marks]**

....

....

....

....

....

....

Turn over for Question 4

Question 4 River landscapes in the UK

Study **Figure 15**, a photograph showing a river in the Lake District at low flow.

Figure 15

0 4 · 1 Identify **one** feature shown in **Figure 15** to provide evidence that the water level in the channel was a lot higher in the past. **[1 mark]**

....

0 4 · 2 Identify **one** feature shown in **Figure 15** to provide evidence that the water level in the channel has **not** been a lot higher for a long time. **[1 mark]**

....

Study **Figure 16**, a photograph of Janet's Foss, a waterfall in North Yorkshire.

0 4 · 3 Define the term 'waterfall'. **[1 mark]**

....

....

....

....

Figure 16

0 4 · 4 Explain the formation of a waterfall and its associated features. **[6 marks]**

Study **Figure 17**, a hydrograph of discharge on Langdon Beck in the Pennines during a storm.

Figure 17

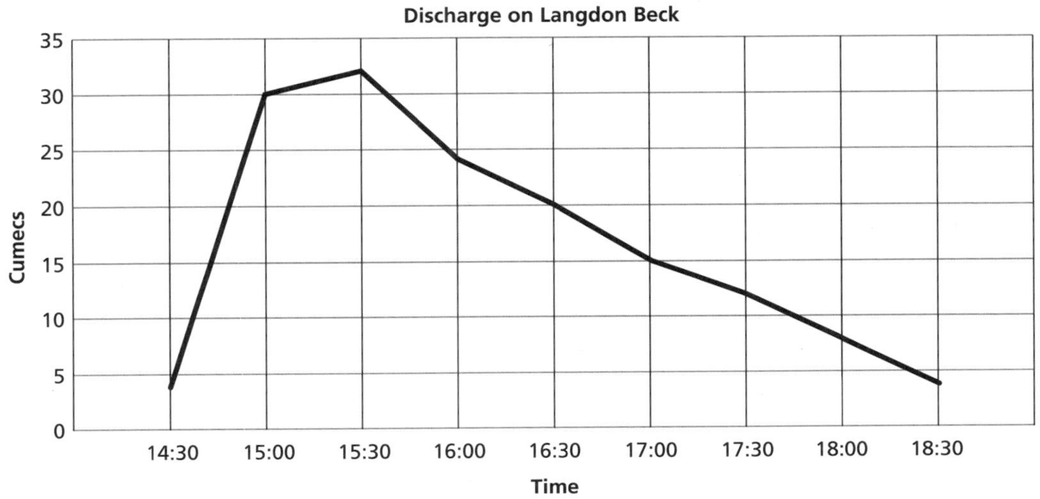

Discharge on Langdon Beck

Cumecs: 0, 5, 10, 15, 20, 25, 30, 35

Time: 14:30, 15:00, 15:30, 16:00, 16:30, 17:00, 17:30, 18:00, 18:30

0 4 · 5 Give the maximum discharge. **[1 mark]**

0 4 · 6 Base flow on Langdon Beck is 6 cumecs.

For how long was discharge greater than base flow? **[1 mark]**

0 4 · 7 Describe some of the effects that the huge increase in discharge might have on the river landscape further downstream. **[4 marks]**

Turn over for Question 5

Question 5 Glacial landscapes in the UK

Study **Figure 18**.

Figure 18

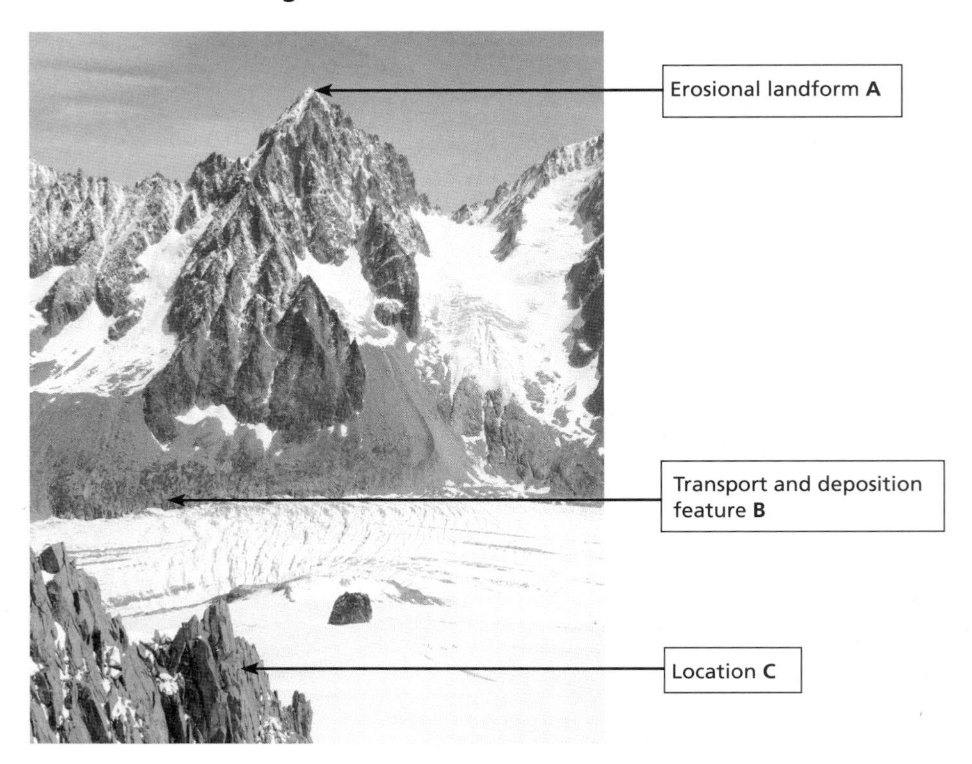

0 5 · 1 Name the erosional landform **A** in **Figure 18**. **[1 mark]**

....

0 5 · 2 Name the transport and deposition feature **B** in **Figure 18**. **[1 mark]**

....

0 5 · 3 Describe the weathering process taking place at location **C** in **Figure 18**. **[2 marks]**

....

....

Study **Figure 19**, a photograph taken from Great Gable in the Lake District.

Figure 19

0 5 · 4 Identify the type of feature labelled **A** in **Figure 19**. **[1 mark]**

....

0 5 · 5 Describe how feature **A** in **Figure 19** has changed over time. **[2 marks]**

....

....

0 5 · 6 Suggest **two** problems faced by farmers in the landscape shown in **Figure 19**. **[2 marks]**

....

....

Question 5 continues on the next page

Study **Figure 20a** and **Figure 20b**, photographs showing the same footpath in the Lake District.

Figure 20a

Figure 20b

0 5 · 7 Describe and explain what has happened to the footpath shown in **Figures 20a** and **20b**. **[6 marks]**

....

....

....

....

....

....

END OF QUESTIONS

Collins

GCSE
GEOGRAPHY

Paper 2 Challenges in the human environment

Time allowed: 1 hour 30 minutes

Materials

For this paper you must have:

- a pencil
- a ruler.

Instructions

- Use black ink or a black ball-point pen.
- Answer **all** questions in Section A and Section B.
- Answer Question 3 and **one** other question in Section C.
- Cross through any work you do not want to be marked.

Information

- The marks for questions are shown in brackets.
- The total number of marks available for this paper is 88.
- Spelling, punctuation, grammar and specialist terminology will be assessed in Question 01.8.

Advice

- For the multiple-choice questions, completely fill in the circle alongside the appropriate answer(s).
  CORRECT METHOD WRONG METHODS
- If you want to change your answer, you must cross out your original answer as shown.
- If you wish to return to an answer previously crossed out, ring the answer you now wish to select as shown.

Name: ....

Practice Exam Paper 2

Section A Urban issues and challenges

Answer **all** questions in this section

Question 1 Urban issues and challenges

Study **Figure 1**, a graph showing the change in the percentage of the global population living in urban and rural areas over a 50-year period.

Figure 1

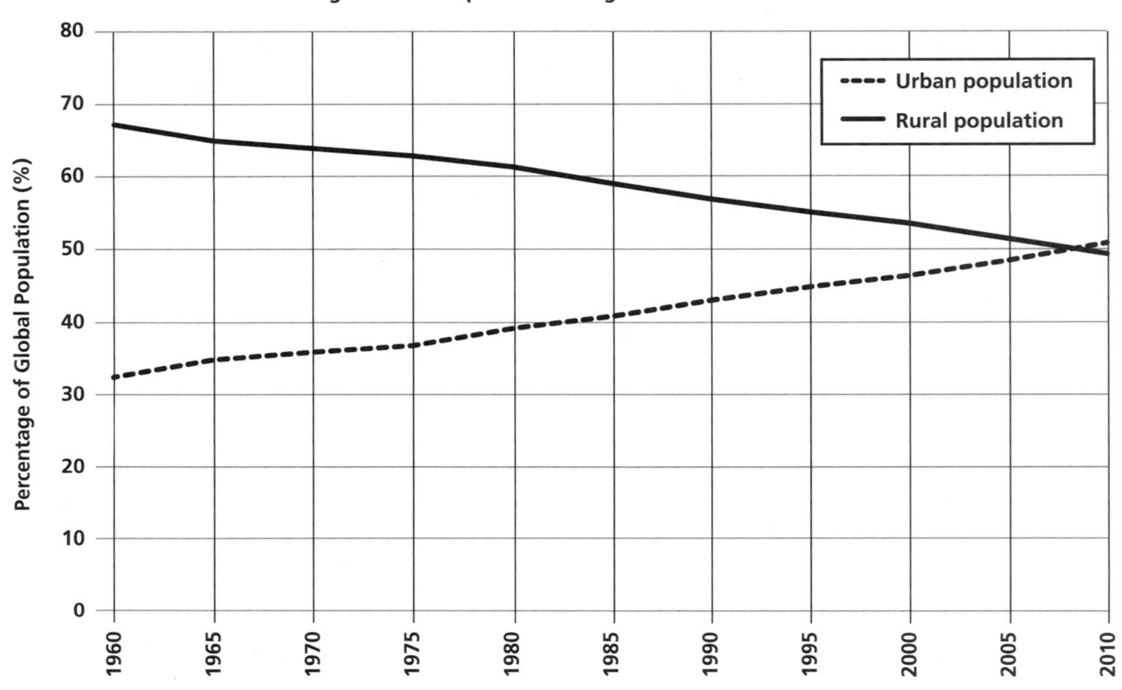

0 1 · 1 Using **Figure 1**, describe **two** changes that occurred to the urban and rural populations between 1960 and 2010. **[2 marks]**

0 1 · 2 Give **two** factors that have led to rapid urbanisation in lower income countries (LICs) and newly emerging economies (NEEs). **[2 marks]**

Study **Figure 2**, a world map showing cities of over five million inhabitants.

Figure 2

Cities of over Five Million Inhabitants

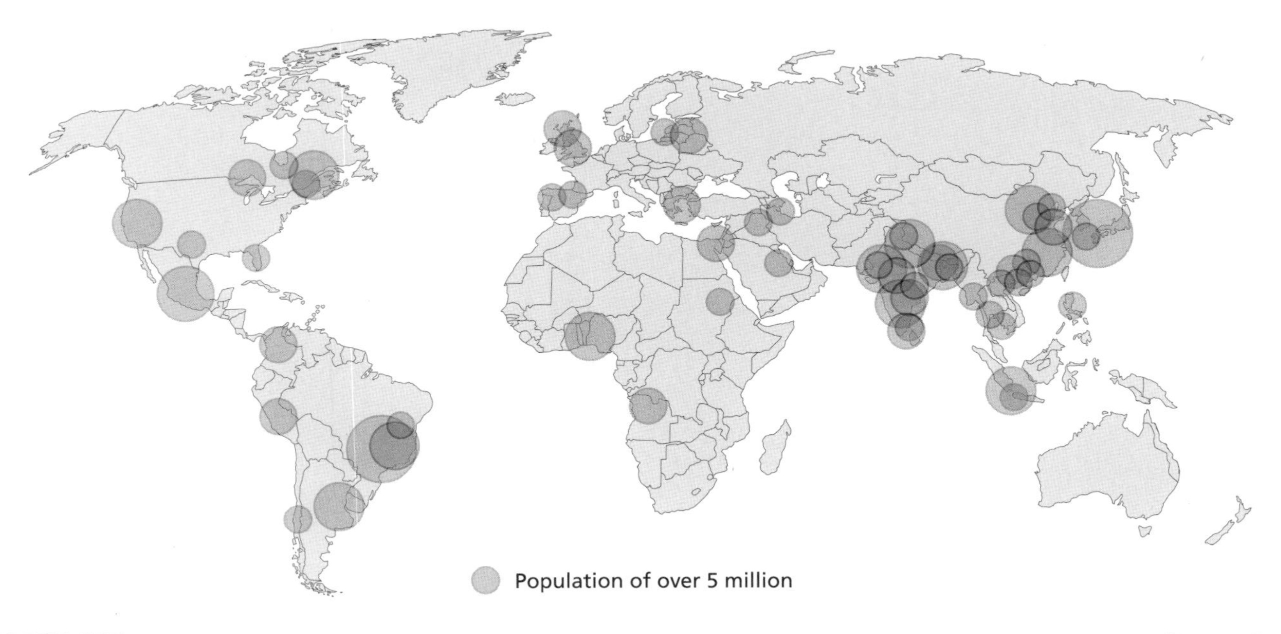

0 1 · 3 Define the term 'mega-city'. **[2 marks]**

....

....

Study **Figure 3**, a photograph showing a slum in Mumbai.

Figure 3

0 1 · 4 Considering the evidence from cities such as Mumbai, explain how growth in cities in LICs or NEEs can create challenges for their populations. **[6 marks]**

....

....

....

....

....

....

....

Question 1 continues on the next page

Study **Figure 4**, an Ordnance Survey map extract of Oxford.

Figure 4

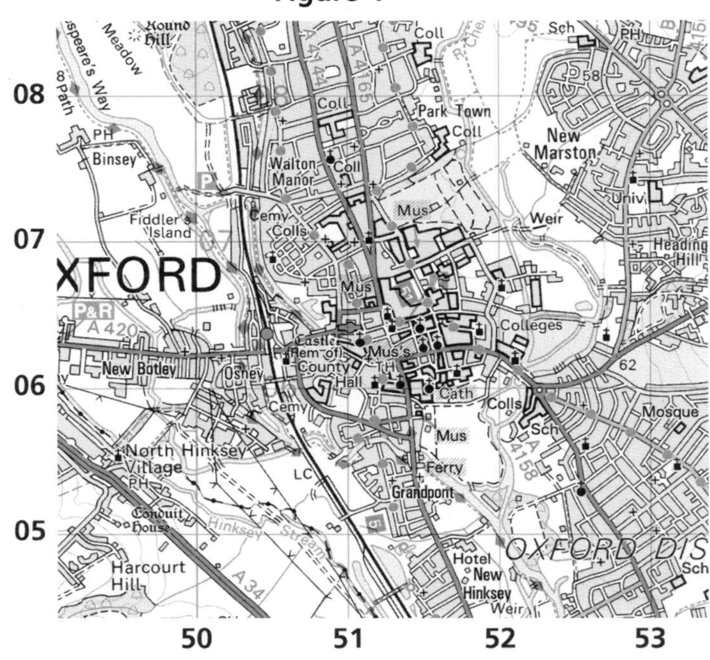

0 1 · 5 Which grid square shows the CBD?

Shade **one** circle only. **[1 mark]**

A 5106 ⬭

B 5005 ⬭

C 5207 ⬭

D 5205 ⬭

E 5105 ⬭

0 1 · 6 Describe the main features of, and explain the need for, a regeneration project in a UK city you have studied. **[6 marks]**

0 1 · 7 Choose from the list **two** features of sustainable urban living.

Shade **two** circles only. **[2 marks]**

A Waste recycling ○

B Creating green space ○

C Large-scale road building ○

D Landfill sites ○

E Building on greenfield sites ○

0 1 · 8 Referring to named examples, explain how urban transport strategies are being used to reduce traffic congestion. **[9 marks] [+ 3 SPaG marks]**

....

....

....

....

....

....

....

....

....

End of Section A

Turn over for Section B

Practice Exam Paper 2

Section B The changing economic world

Answer **all** questions in this section

Question 2 The changing economic world

0 2 · 1 Define the term 'birth rate'. **[2 marks]**

0 2 · 2 Apart from birth rate, give **three** other social measures of development. **[3 marks]**

0 2 · 3 Explain the limitations of using economic indicators as a means to measure development. **[4 marks]**

0 2 · 4 Give **two** strategies for narrowing the development gap. **[2 marks]**

Study **Figure 5**, a family planning advertisement from rural India.

Figure 5

0 2 · 5 Describe how family planning programmes, such as those in India, can help to narrow the development gap. **[4 marks]**

....

....

....

....

Study **Figure 6**, a map of Silicon Roundabout in London.

Figure 6

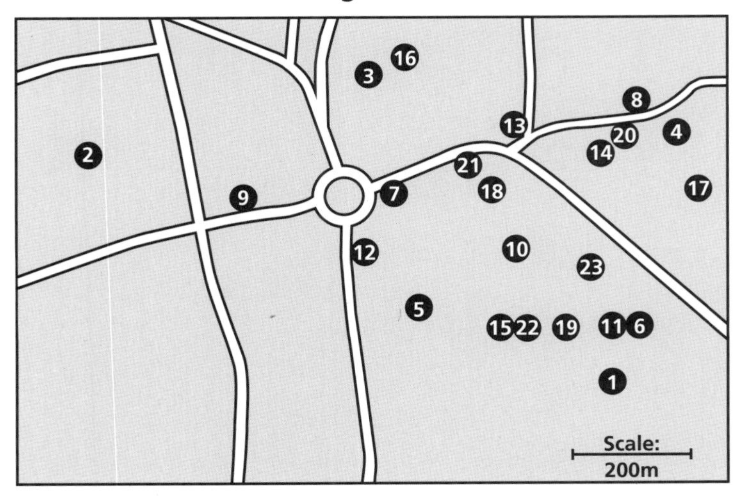

Question 2 continues on the next page

0 2 · 6 Using **Figure 6**, measure the direct distance in metres between businesses 5 and 14. **[1 mark]**

....

0 2 · 7 Places such as Silicon Roundabout in London and Silicon Fen in Cambridge have seen huge recent growth in high-tech industry.

Suggest **three** features that a high-tech company would be looking for when choosing where to locate its facilities. **[3 marks]**

....

....

....

0 2 · 8 With reference to a case study of one LIC or NEE:

Explain how economic development is improving the quality of life for the population of a place you have studied. **[9 marks]**

....

....

....

....

....

....

....

....

....

0 2 · 9 Give **two** ways in which modern industrial development in more developed countries, such as the UK, can be made more sustainable. **[2 marks]**

....

....

End of Section B

Section C The challenge of resource management

Answer Question 3 and **either** Question 4 **or** Question 5 **or** Question 6

Question 3 The challenge of resource management

Study **Figures 7** and **8**, showing information about the number of people with and without access to clean drinking water in different parts of the world.

Figure 7

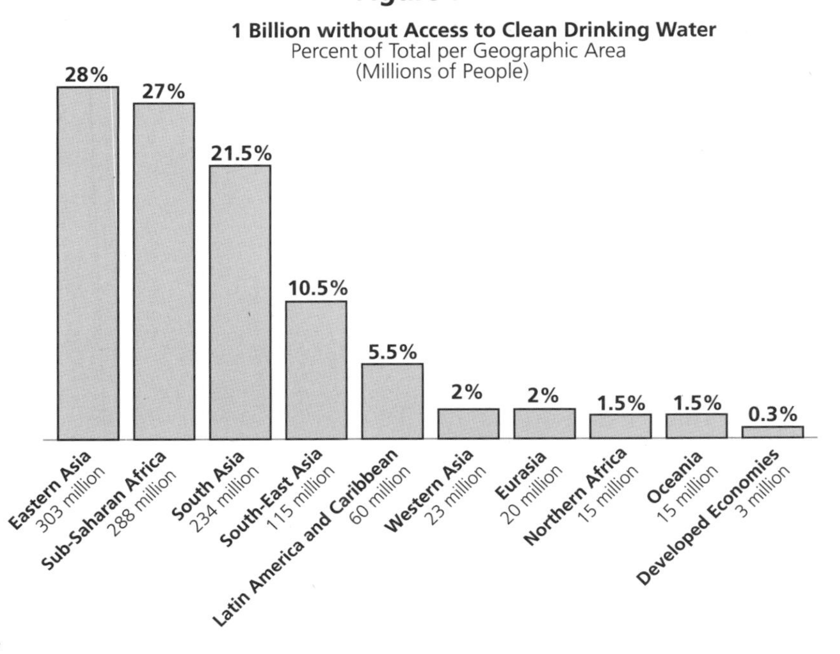

Figure 8

Global Water Availability

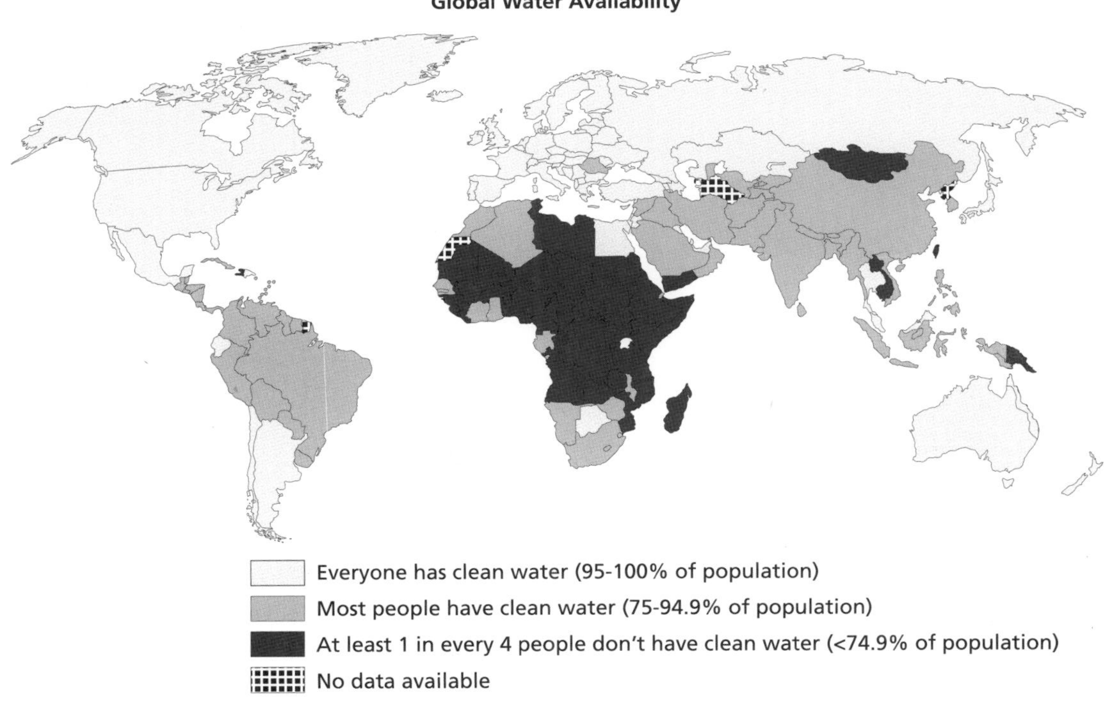

Practice Exam Paper 2

0 3 · 1 Using **Figures 7** and **8**, identify the types of country that have less access to clean water.

Tick **one** option only. **[1 mark]**

A Higher income countries (HICs) ☐

B Lower income countries (LICs) and newly emerging economies (NEEs) ☐

C European Union (EU) countries ☐

D Developed economies ☐

0 3 · 2 Suggest **two** reasons why a country may have less access to clean water. **[2 marks]**

....

....

0 3 · 3 Using examples, explain why water must sometimes be transferred between areas to maintain supplies. **[6 marks]**

....

....

....

....

....

....

Study **Figure 9**, a graph showing global consumption of energy from different sources since 1990.

Figure 9

Growth in Fossil Fuel, Wind and Solar Consumption

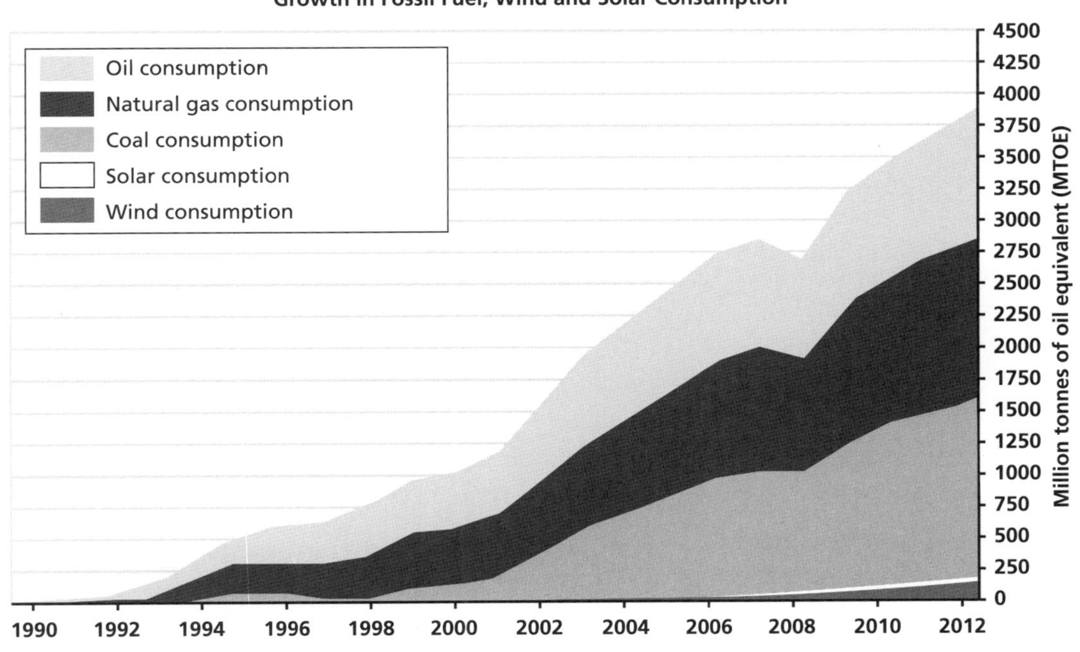

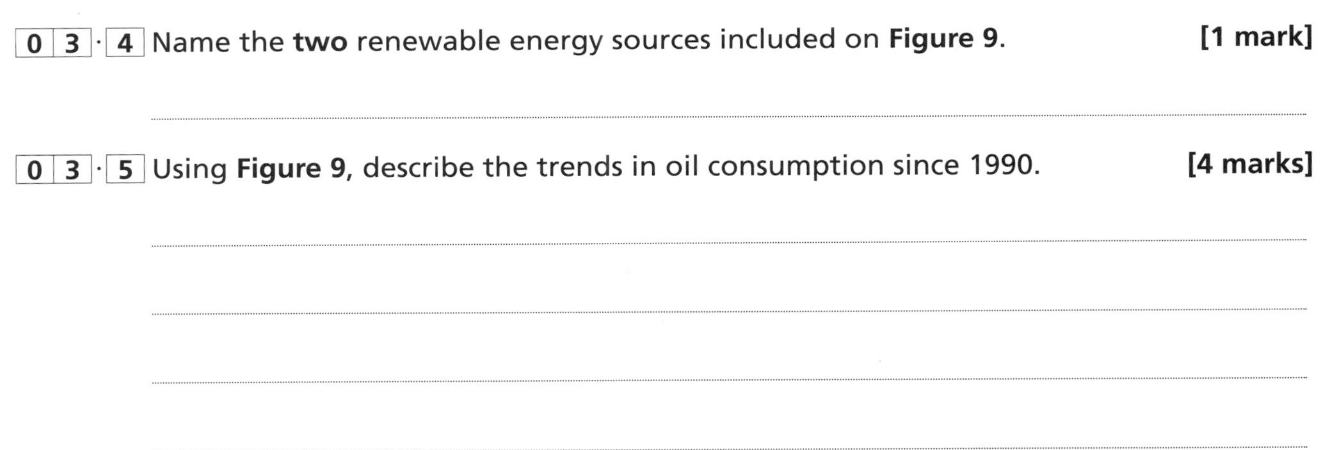

0 3 · 4 Name the **two** renewable energy sources included on **Figure 9**. **[1 mark]**

0 3 · 5 Using **Figure 9**, describe the trends in oil consumption since 1990. **[4 marks]**

Practice Exam Paper 2

Answer **either** Question 4 (Food) **or** Question 5 (Water) **or** Question 6 (Energy)

Question 4 Food

Study **Figures 10** and **11**, charts showing a breakdown of the average daily calorific intake for the UK and Somalia by food type.

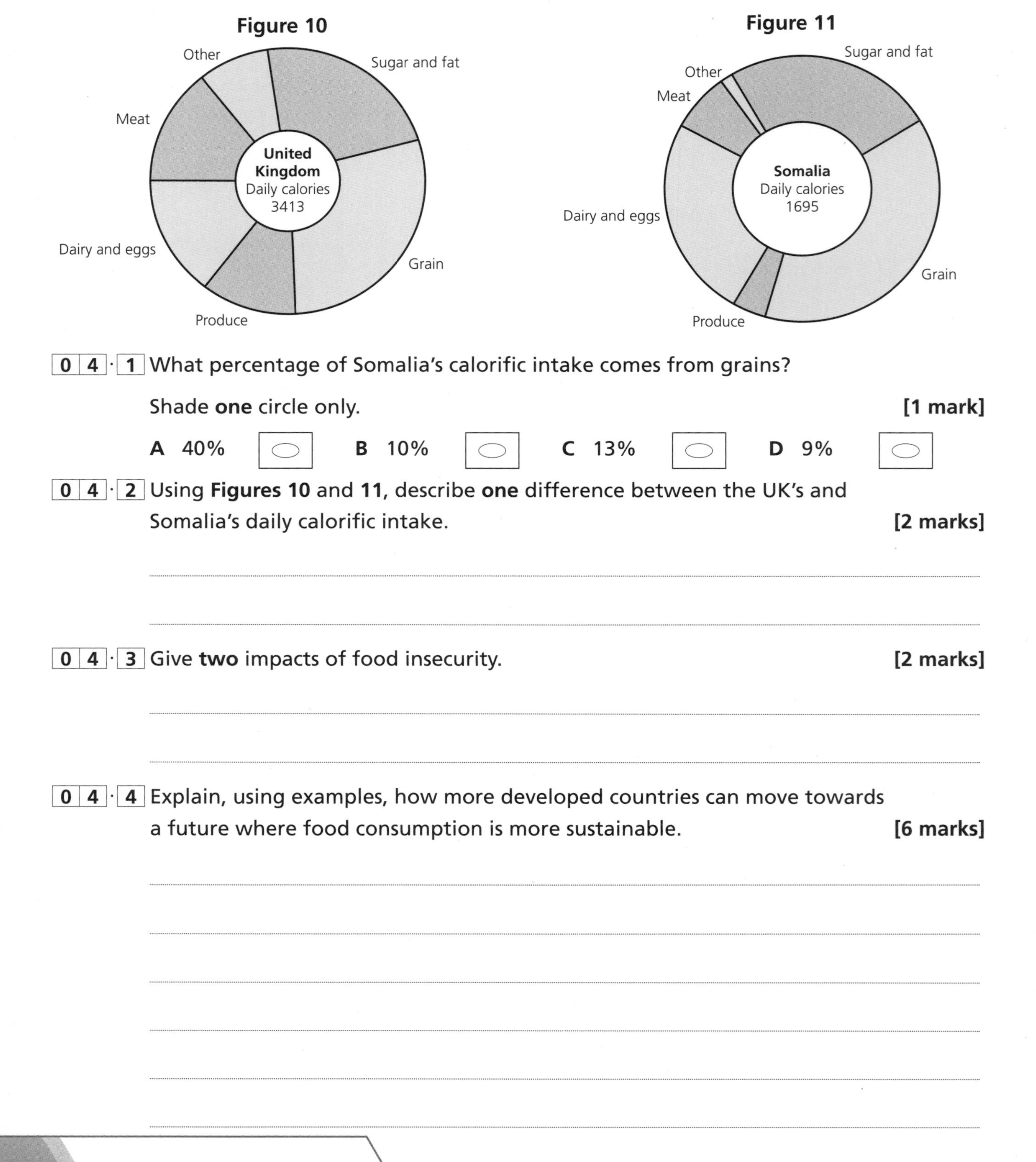

0 4 · 1 What percentage of Somalia's calorific intake comes from grains?

Shade **one** circle only. **[1 mark]**

A 40% ◯ **B** 10% ◯ **C** 13% ◯ **D** 9% ◯

0 4 · 2 Using **Figures 10** and **11**, describe **one** difference between the UK's and Somalia's daily calorific intake. **[2 marks]**

....

....

0 4 · 3 Give **two** impacts of food insecurity. **[2 marks]**

....

....

0 4 · 4 Explain, using examples, how more developed countries can move towards a future where food consumption is more sustainable. **[6 marks]**

....

....

....

....

....

....

Question 5 Water

Study **Figure 12**, a graph showing average daily water use per person by country.

Figure 12

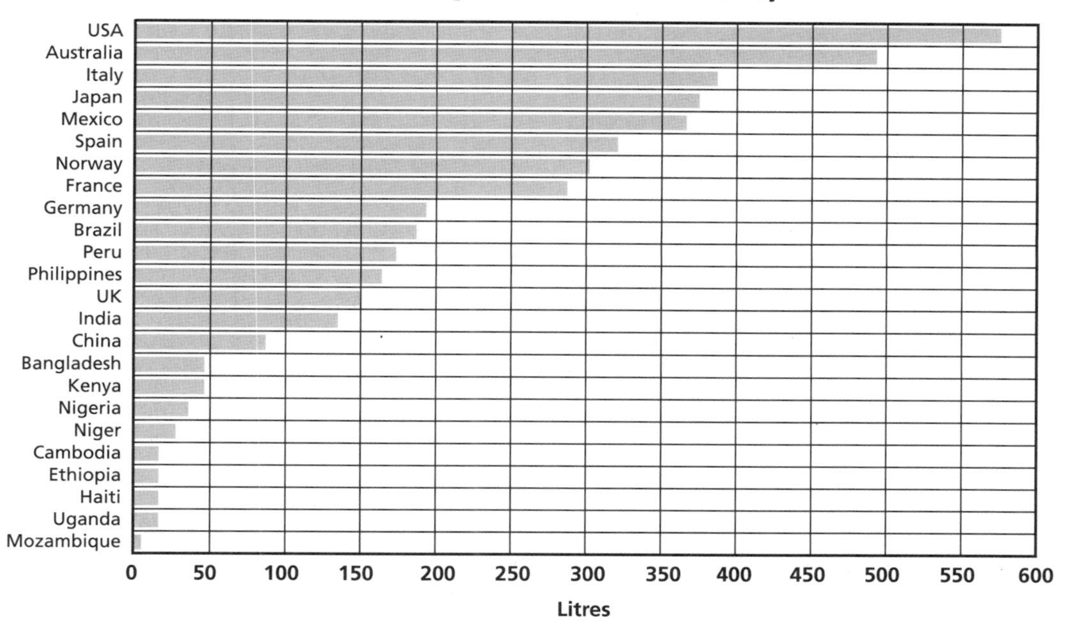

0 5 · 1 What is the average amount of water used per person per day in Kenya?

Shade **one** circle only. **[1 mark]**

A Less than 40 litres per person per day ⬭

B 40–50 litres per person per day ⬭

C 100–120 litres per person per day ⬭

D 75–85 litres per person per day ⬭

0 5 · 2 Using data from **Figure 12**, describe the differences in water consumption between the USA and China. **[2 marks]**

....

....

0 5 · 3 Give **two** impacts of water insecurity. **[2 marks]**

....

....

Turn over

0 5 · 4 Explain how water resources could be used more sustainably in the future. **[6 marks]**

Question 6 Energy

Study **Figure 13**, a map showing the approximate installed capacity and actual amount of energy delivered by hydroelectric power stations in 2013, compared to the potential for generating hydroelectric power, in six regions of the world.

Figure 13

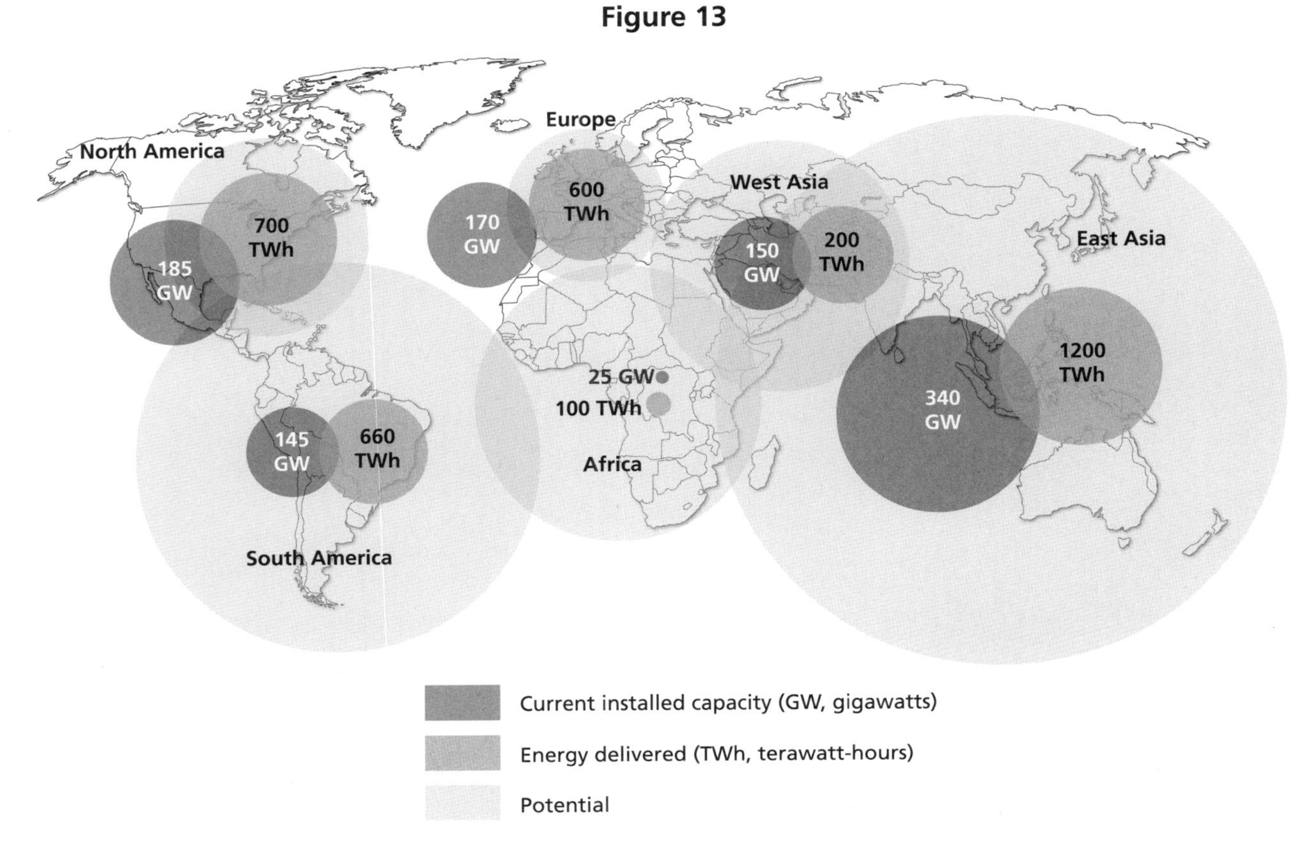

0 6 · 1 Using **Figure 13**, which region has the greatest potential for generating hydroelectric power?

Shade **one** circle only. **[1 mark]**

A North America ○

B South America ○

C Africa ○

D Europe ○

E East Asia ○

F West Asia ○

Question 6 continues on the next page

0 6 · 2 Suggest **two** reasons why the installed hydroelectric power capacity in Africa was significantly lower than in the other five regions. **[2 marks]**

0 6 · 3 Give **two** impacts of energy insecurity. **[2 marks]**

0 6 · 4 Using examples, explain how the extraction of a fossil fuel can have both positive and negative effects. **[6 marks]**

END OF QUESTIONS

Collins

GCSE
GEOGRAPHY

Paper 3 Geographical applications

Time allowed: 1 hour 15 minutes

Materials

For this paper you must have:

- a clean copy of the pre-release resources booklet (see page 218).

Instructions

- Use black ink or a black ball-point pen.
- Answer all questions.
- Cross through any work you do not want to be marked.

Information

- The marks for questions are shown in brackets.
- The total number of marks available for this paper is 76.
- Spelling, punctuation and grammar will be assessed in Questions 03.2 and 05.6.

Advice

- For the multiple-choice questions, completely fill in the circle alongside the appropriate answer(s).

  CORRECT METHOD   WRONG METHODS

- If you want to change your answer, you must cross out your original answer as shown.
- If you wish to return to an answer previously crossed out, ring the answer you now wish to select as shown.

Make sure you have studied the resources booklet before completing this practice paper.

Name: ....

Practice Exam Paper 3

Section A Issue Evaluation

Answer **all** questions in this section

Question 1 Issue evaluation

Study **Figure 1**, a world map showing the distribution of tropical rainforests.

Figure 1

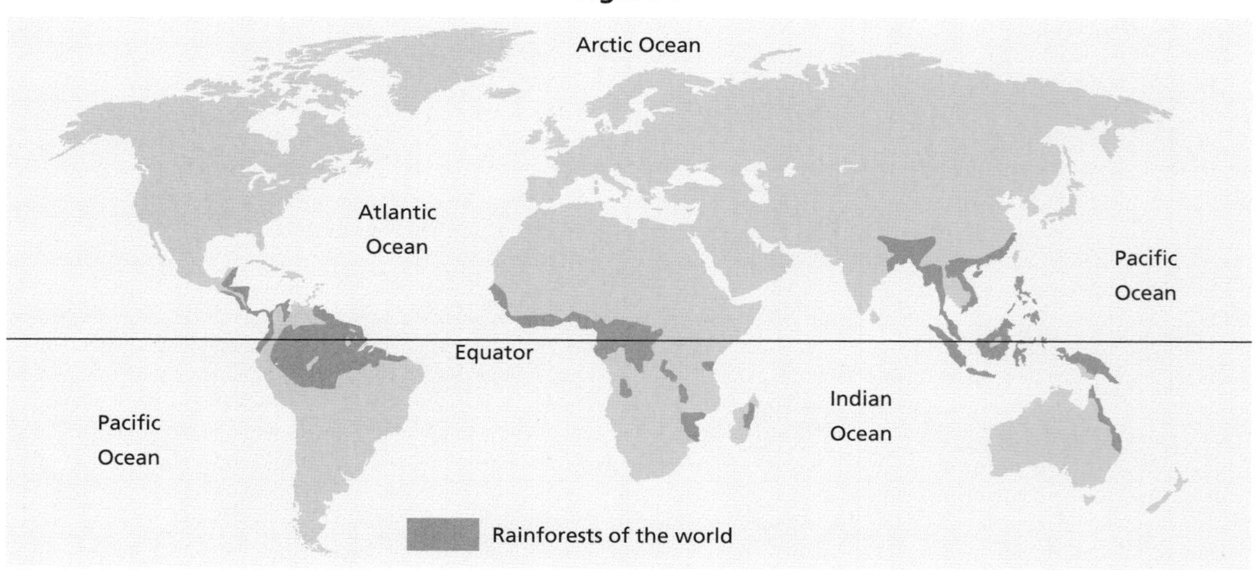

0 1 · 1 Using **Figure 1**, which **two** of the following statements are true?

Shade **two** circles only. **[2 marks]**

A Rainforests are found mainly in the Southern Hemisphere. ⬭

B There is a band of rainforests north and south of the Equator. ⬭

C The largest area of rainforest is in Africa. ⬭

D There are no rainforests in south-east Asia. ⬭

E The largest area of rainforest is in South America. ⬭

Study **Figure 2**, a table showing the 10 countries that have experienced the greatest loss of rainforest (in hectares) in the 21st century so far.

Figure 2

Country	**Rank**	**Average loss 2010–2014**	**Trend**
Brazil	1	2 347 727	Down
Indonesia	2	1 543 623	Up
DR Congo	3	778 348	Up
Malaysia	4	469 511	Up
Paraguay	5	406 785	Up
Bolivia	6	291 167	Down
Myanmar	7	207 677	Up
Madagascar	8	203 165	Up
Cambodia	9	187 893	Up
Peru	10	187 196	Up

0 1 · 2 Describe **two** links between the locations of the countries and the patterns shown in the chart. **[2 marks]**

Study **Figure 3**, 'Threats to Rainforests and Biodiversity and their Impacts', in the resources booklet.

0 1 · 3 Explain why rainforests have such high levels of biodiversity. **[6 marks]**

Turn over for Question 2

Practice Exam Paper 3

Study **Figure 4**, a graph showing mean global temperature changes since 1880.

Figure 4

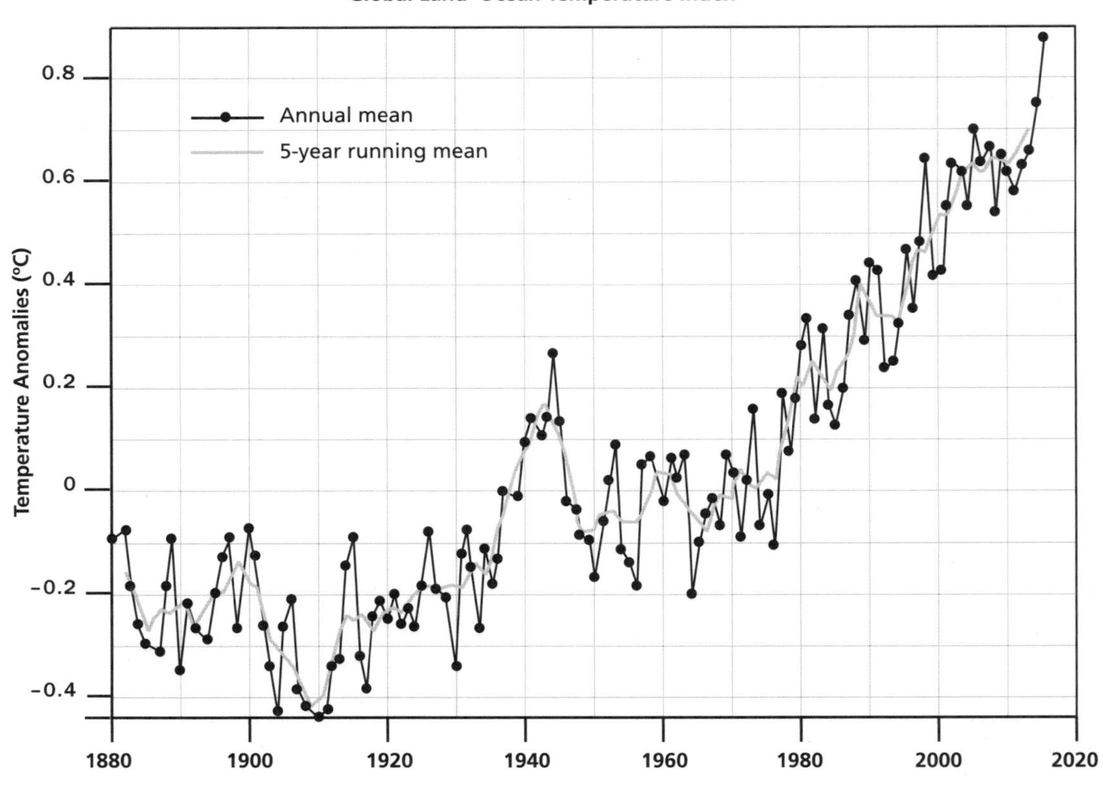

0 2 · 1 Identify the general trend shown in **Figure 4**. **[1 mark]**

....

0 2 · 2 What was the difference in mean temperature between 1910 and 2010? **[1 mark]**

....

0 2 · 3 In which decade was mean temperature change greatest? **[1 mark]**

....

Study **Figure 5a**, a map showing the predicted change in GDP per capita by 2100 compared to a world without climate change, and **Figure 5b**, a table showing GDP per capita in 2014.

Figure 5a

Estimated Economic Impact of Climate Change by 2100

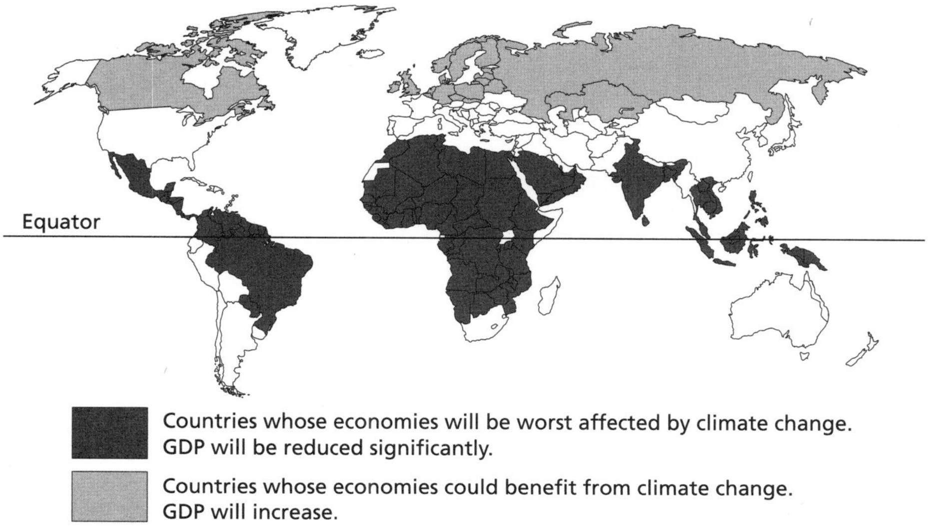

Equator

Countries whose economies will be worst affected by climate change. GDP will be reduced significantly.

Countries whose economies could benefit from climate change. GDP will increase.

Figure 5b

Country	GDP per capita 2014 ($US)
Brazil	11 124
DR Congo	311
Ecuador	4657
Indonesia	3125
Malaysia	9069
UK	38 292

0 2 · 4 Discuss possible links between rainforest destruction, climate change and the economic information given in **Figures 5a** and **5b**. **[6 marks]**

....

....

....

....

....

....

Turn over for Question 3

0 3 · 1 Using **Figure 6**, 'The Challenges of Rainforest Conservation', in the resources booklet and your own knowledge and understanding:

Discuss some of the conflicts that could arise over rainforests and their exploitation between the richer countries of the world and the countries where rainforests are found. **[6 marks]**

....

....

....

....

....

....

There are many organisations which work to conserve the rainforests. They have developed a range of different conservation strategies to benefit the whole world, which also take into account the needs of local people and economic issues in the countries where the rainforests are found.

Study **Figure 7a** and **Figure 7b**, statements by two such organisations that need to raise awareness and money.

Figure 7a

Organisation A

We are an NGO that works with local NGO conservation groups rather than governments. With donations from people around the world, we buy rainforest land to create reserves, which are passed on to the local groups.

We give scientific or technical advice and employ 'Rangers' who have a variety of roles: monitoring their area; keeping trails clear; maintaining fences; carrying out repair work after fires or storms; growing and planting trees; giving education talks; leading tourist walks; and assisting researchers. The presence of these Rangers deters poaching and other illegal activity and they earn a wage, which gives them independence and status within their community.

Our projects are small-scale but sustainable. We aim to protect the world's most biologically important and threatened habitats acre by acre.

Since being founded in 1989, we have funded partner organisations around the world to create reserves and give permanent protection to habitats and wildlife.

Figure 7b

Organisation B

We are an NGO involved in conservation work. We work with governments, providing expertise and knowledge to help maintain protected areas that they have created. A typical scheme is working with individual native tribes in their own rainforest territory.

Donations from around the world provide boats, radios, fuel, border control training and aerial surveys. Surveillance has stopped illegal mining and mapped locations vulnerable to logging and fishing.

We are setting up non-timber, sustainable businesses with nuts, fruit, honey and essential oils. These provide jobs in harvesting, processing and transport.

We are dedicated to managing the things that we can control. We want societies to responsibly and sustainably care for nature and our global biodiversity, for the well-being of all humanity.

0 3 · 2 Which of the two organisations do you think will most effectively address the issues arising from threats to the rainforest?

Use evidence from the resources booklet and your own understanding to explain why you have reached the decision. **[9 marks] [+ 3 SPaG marks]**

Turn over for Question 4

Section B Fieldwork

Answer **all** questions

0 4 · 1 In the context of fieldwork, what do you understand by the term 'sampling'? **[1 mark]**

....

0 4 · 2 What is a line sample? **[1 mark]**

....

0 4 · 3 A school group is carrying out a questionnaire survey in a town centre about shopping.

Explain which type of sampling would be most appropriate and reliable. **[2 marks]**

....

....

Study **Figure 8**, a pie chart showing the results of a questionnaire survey about the type of housing in which respondents live.

Figure 8

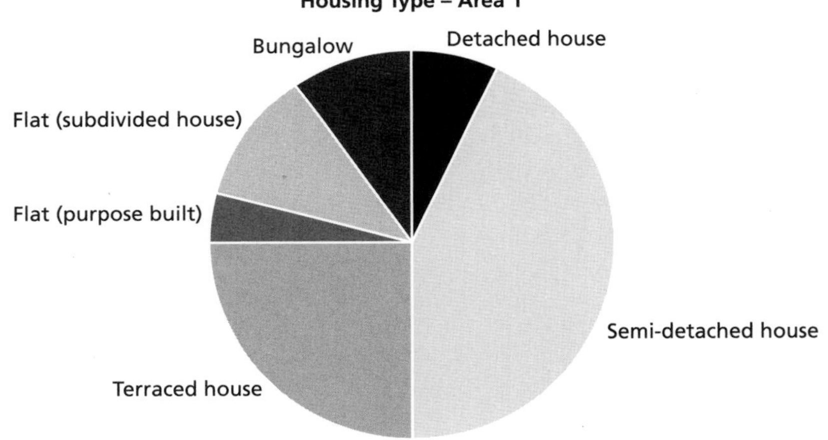

0 4 · 4 Estimate the percentage of respondents from Area 1 living in terraced housing. **[1 mark]**

....

Study **Figure 9a**, a table showing the results of a similar survey carried out in a different part of the same town, and **Figure 9b**, a divided bar chart produced by a student to show the results.

Figure 9a

Housing type – Area 2	%
Detached house	1
Semi-detached house	23
Terraced house	46
Flat (purpose built)	10
Flat (subdivided house)	20
Bungalow	0

Figure 9b

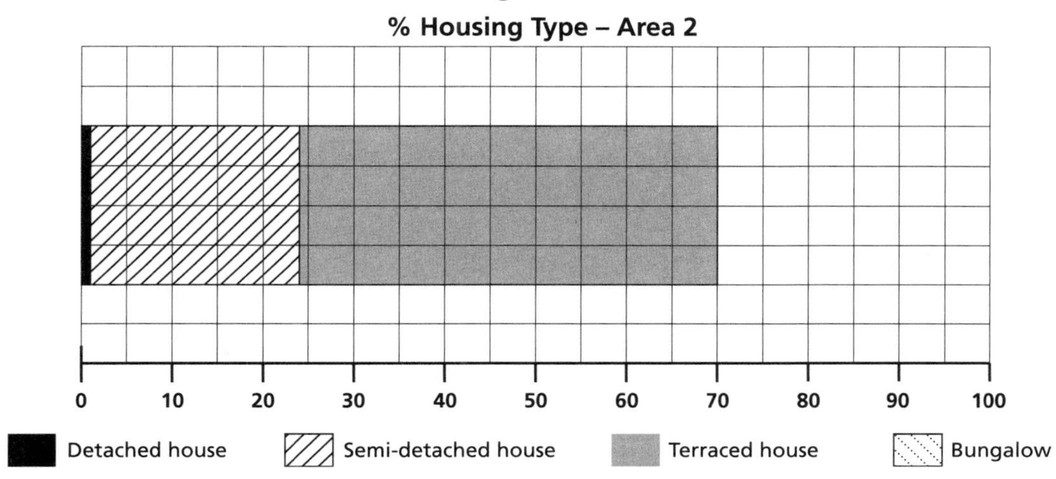

0 4 · 5 Complete **Figure 9b** to show the percentage of respondents living in the **two** different types of flat. **[2 marks]**

0 4 · 6 What is the modal class of housing type in Area 2? **[1 mark]**

....

Question 4 continues on the next page

Practice Exam Paper 3

Study **Figure 10**, a diagram showing visits by people living in villages to nearby towns to shop for clothes.

Figure 10

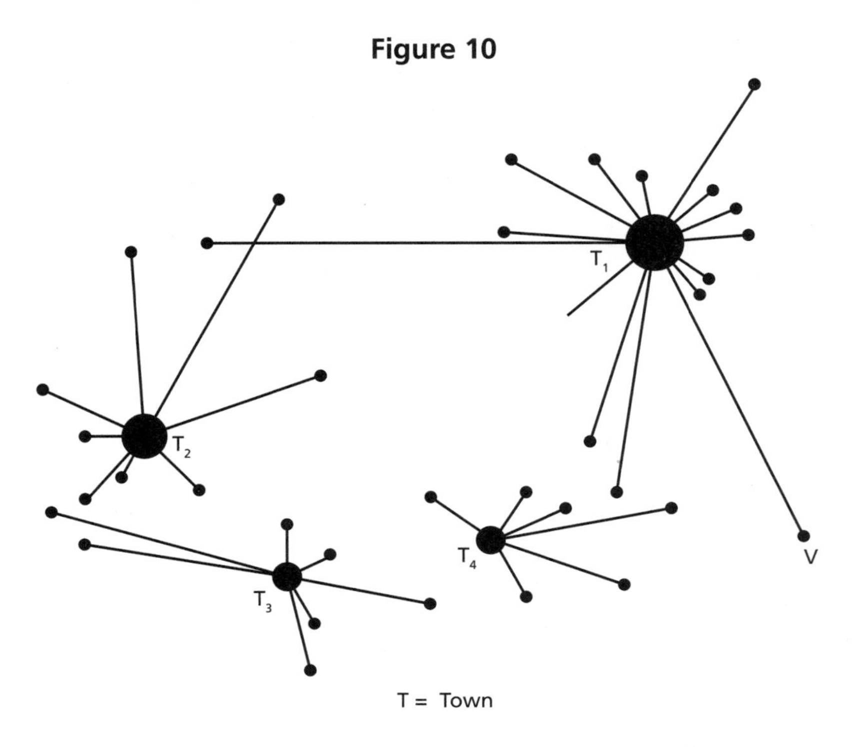

0 4 · 7 Name the type of diagram shown in **Figure 10**. **[1 mark]**

....

0 4 · 8 Suggest **one** reason why people from village V might travel to town T_1. **[1 mark]**

....

04·9 A group of students is measuring bedload on a moorland river.

They are interested in how the long axis of bedload changes downstream so take 25 samples across the channel width at 10 locations starting from near the source. Their results are shown in the table below.

Location	Long Axis of Longest Sample (mm)	Long Axis of Shortest Sample (mm)	Mean Length of Long Axis of all 25 Samples (mm)
1	360	94	250
2	275	78	176
3	343	81	201
4	212	55	153
5	246	37	131
6	197	32	53
7	148	28	51
8	112	14	29
9	73	11	17
10	44	5	8

The students want to present their results to show all three pieces of data at each location.

Draw a diagram to show how the information might be presented. You should not plot the figures. **[3 marks]**

Question 4 continues on the next page

Practice Exam Paper 3

Study **Figure 11**, a map showing travel times by road (in minutes) to a large town from a number of locations in the surrounding area.

The 5, 10 and 25-minute isochrones (isolines joining up places of equal time) have been inserted.

Figure 11

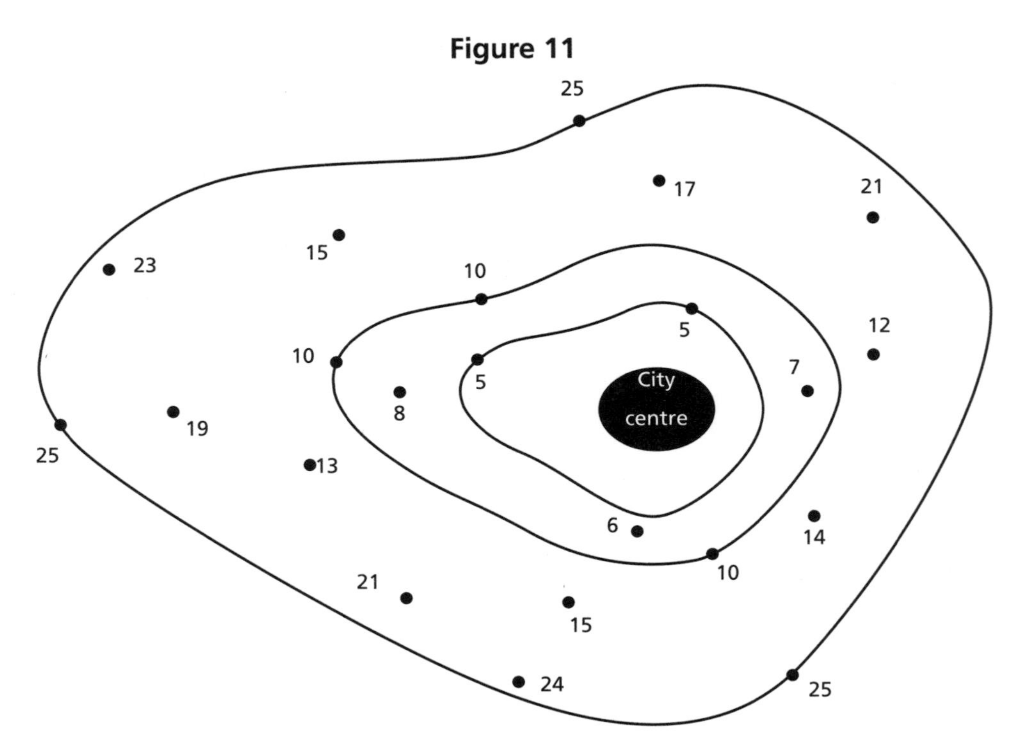

0 4 · 1 0 Add the isochrones for 15 minutes and 20 minutes. **[2 marks]**

0 4 · 1 1 **Figure 11** shows that it is much quicker to travel into the town from some directions than others.

From which direction are the slowest routes?

Shade **one** circle only. **[1 mark]**

A north east ○

B south east ○

C south west ○

D north west ○

State the title of your fieldwork enquiry in which **physical** geography data were collected.

0 5 · 1 Describe how you used a piece of equipment in your physical geography enquiry. **[2 marks]**

0 5 · 2 Justify a **presentation** technique used in analysing data for your physical geography enquiry. **[3 marks]**

0 5 · 3 Justify a **statistical** method you used in analysing data for your physical geography enquiry. **[3 marks]**

State the title of your fieldwork enquiry in which **human** geography data were collected.

0 5 · 4 Describe **one** possible risk in carrying out your enquiry. **[1 mark]**

0 5 · 5 Explain how this risk was reduced. **[2 marks]**

0 5 · 6 What conclusions did your results enable you to make? To what extent were these in line with the geographical concepts or theories underpinning the enquiry? **[9 marks] [+ 3 SPaG marks]**

END OF QUESTIONS

Collins

GCSE GEOGRAPHY

Resources for Paper 3 Geographical applications

Study the resources in this booklet before completing Practice Paper 3.

The resources for Paper 3 of your actual exam will be issued to you 12 weeks before the date of the exam.

This booklet contains two resources as follows:

Figure 3

Threats to Rainforests and Biodiversity and their Impacts

Distribution of Tropical Rainforests

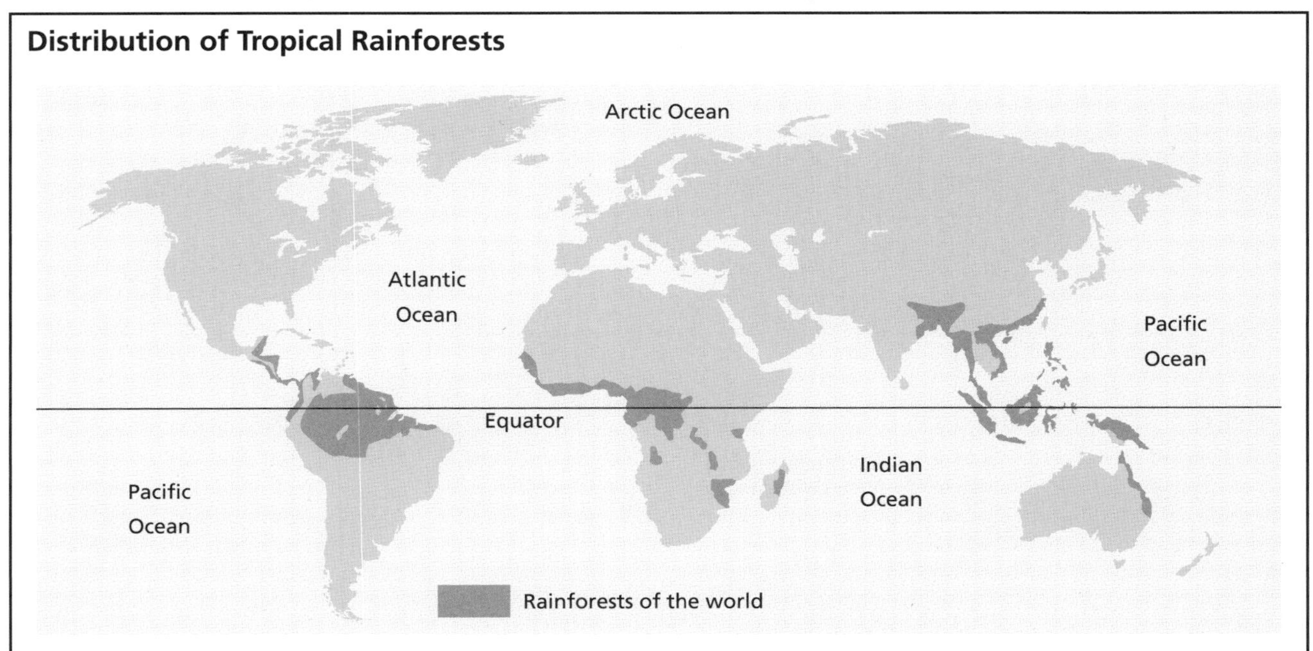

Biodiversity

Biodiversity generally refers to the range of species of plants and animals in a region or country. Countries containing tropical rainforest score more highly for biodiversity than those without. Indonesia is the most biodiverse country in the world.

Biodiversity Scores by Country

Country	Biodiversity score
Indonesia	1.00
Colombia	0.94
Mexico	0.93
Brazil	0.88
Ecuador	0.88
Iceland	0.10

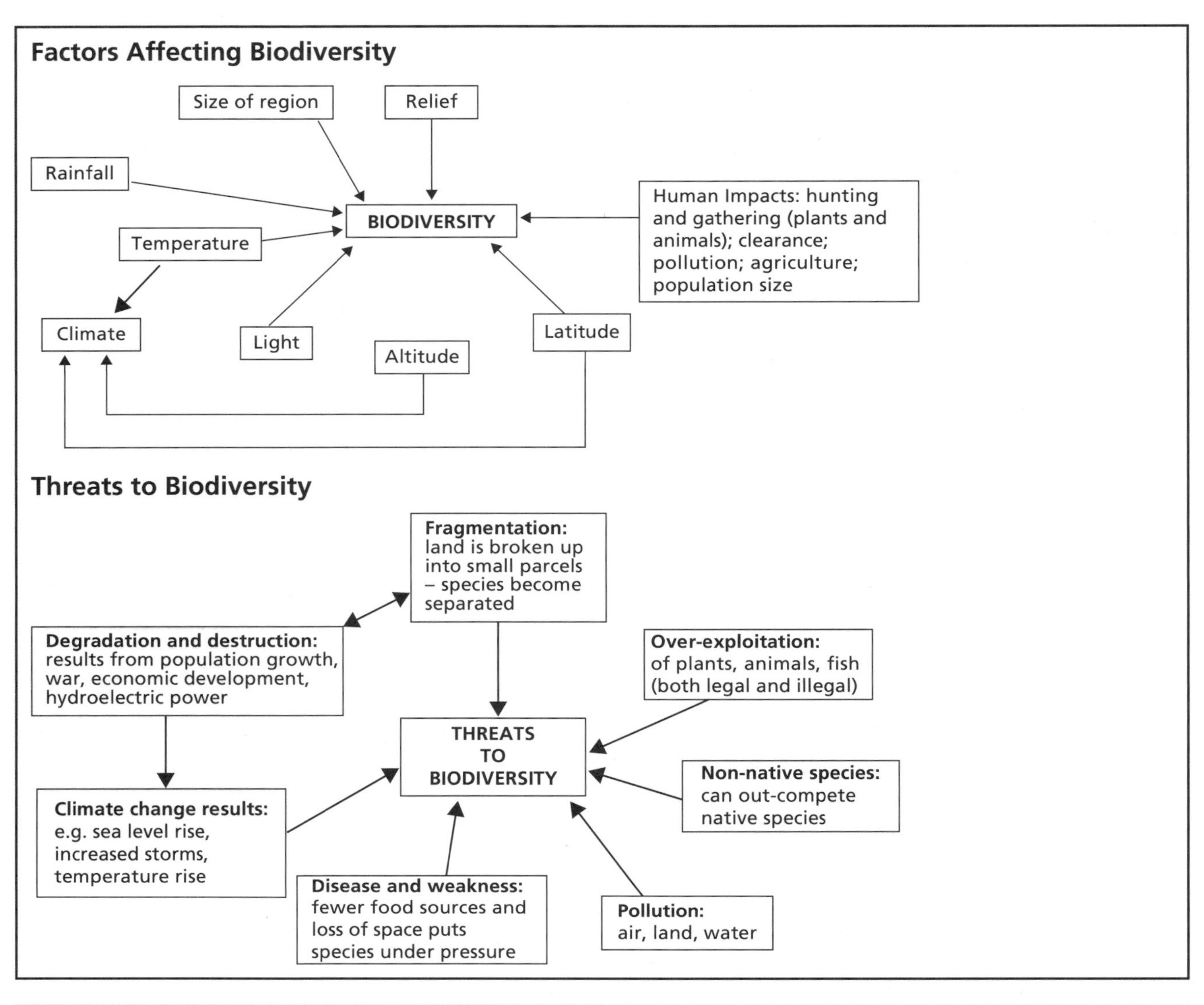

Fragmentation of Forest as Logging Takes Place

Estimates of Tropical Forest Loss (hectares)

Country	Rank	Average loss 2010–2014	Trend
Brazil	1	2 347 727	Down
Indonesia	2	1 543 623	Up
DR Congo	3	778 348	Up
Malaysia	4	469 511	Up
Paraguay	5	406 785	Up
Bolivia	6	291 167	Down
Myanmar	7	207 677	Up
Madagascar	8	203 165	Up
Cambodia	9	187 893	Up
Peru	10	187 196	Up

Empty Forests

More than half of all tropical protected areas may be 'empty forests' — containing trees but few animals as a result of over-exploitation and uncontrolled hunting. As a result, animal species are in danger of extinction, tree species lose important seed dispersal and local people lose an important supply of protein.

Tree Diversity on Rainforest Alliance Certified Coffee Farms in El Salvador

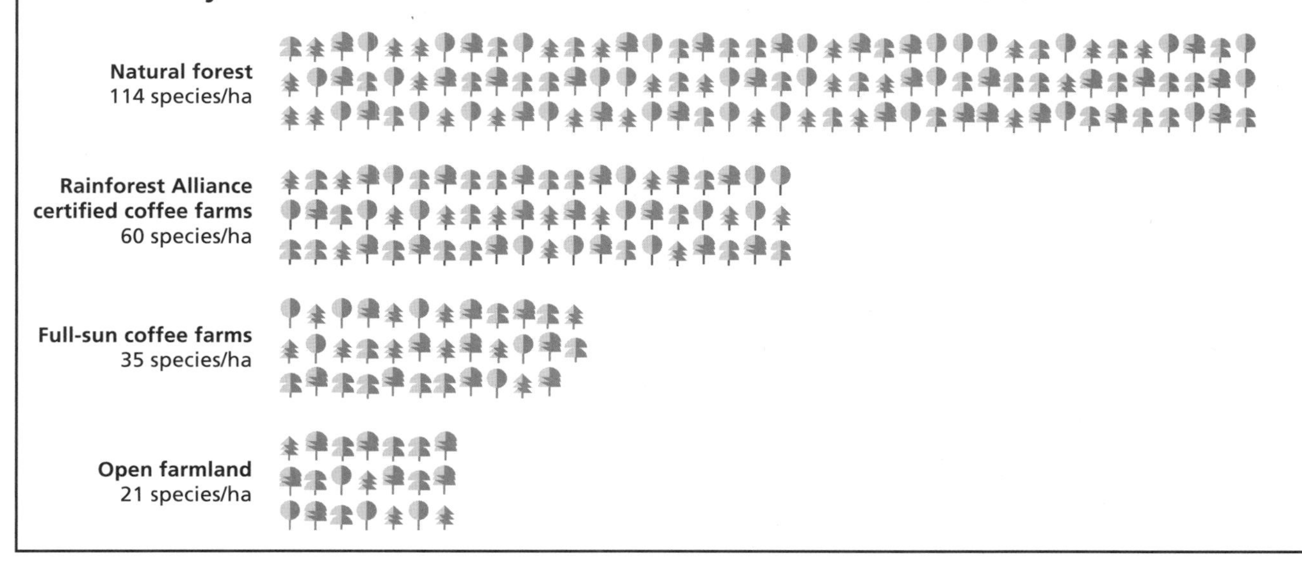

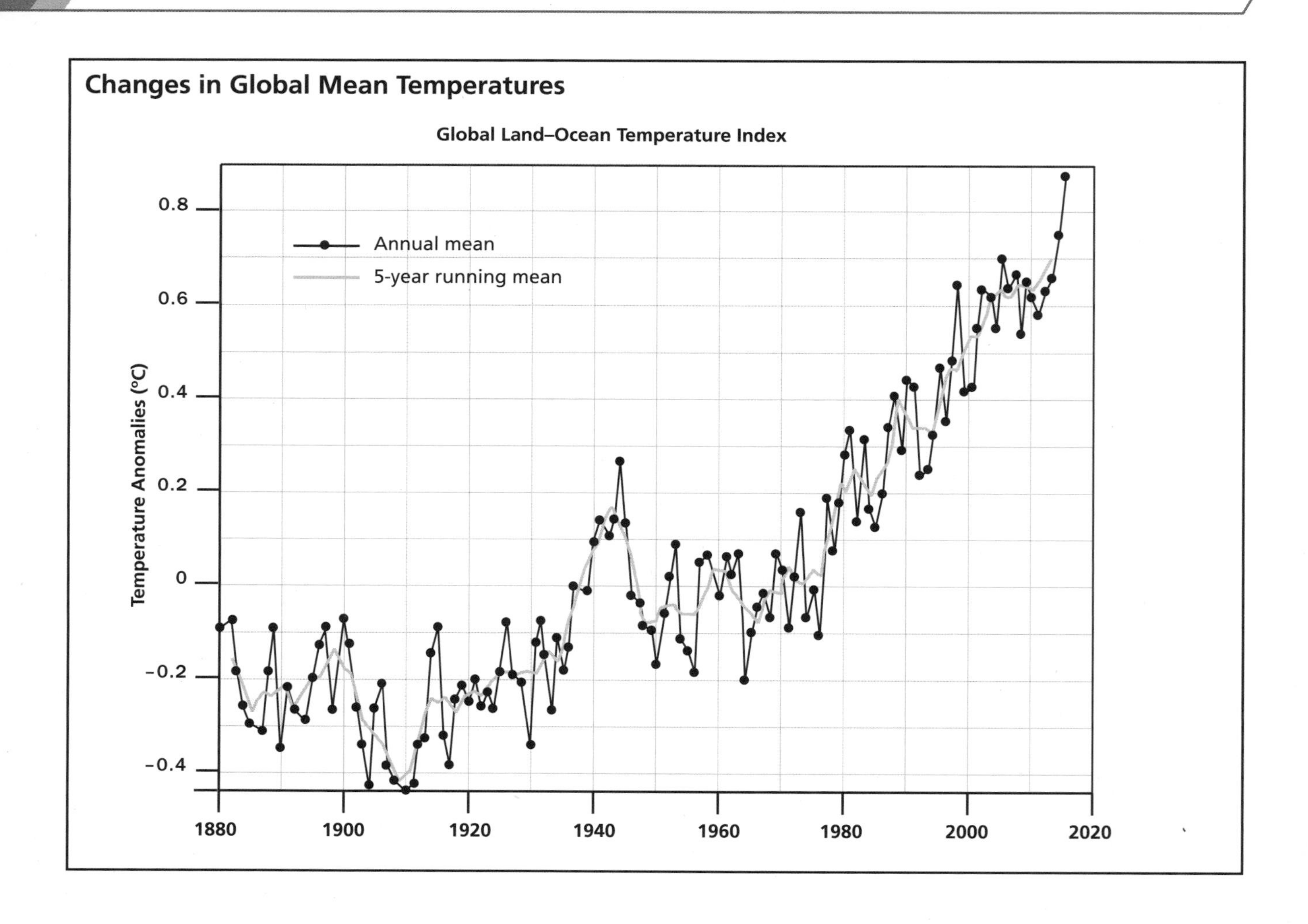
Changes in Global Mean Temperatures
Global Land–Ocean Temperature Index
Annual mean
5-year running mean
Temperature Anomalies (°C)
0.8
0.6
0.4
0.2
0
-0.2
-0.4
1880
1900
1920
1940
1960
1980
2000
2020

Climate Change and GDP Prediction

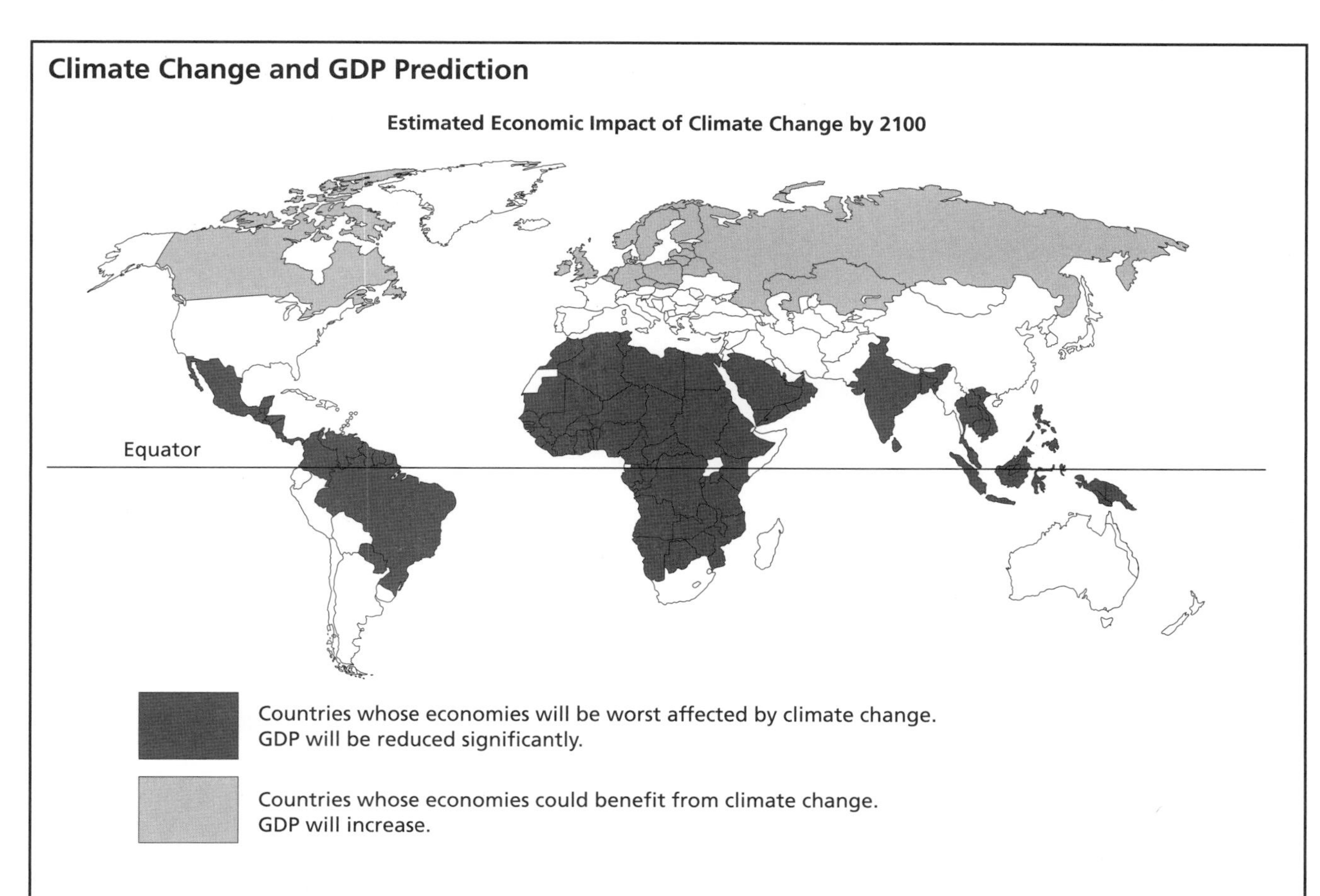

GDP per Capita in 2014 ($US) by Country

Country	GDP per capita 2014 ($US)
Brazil	11 124
DR Congo	311
Ecuador	4657
Indonesia	3125
Malaysia	9069
UK	38 292

Figure 6

The Challenges of Rainforest Conservation

The Value of the Forests

Tropical forests provide a range of goods and invaluable services. They are very important to the hydrological cycle (i.e. rain and water systems), and they maintain some of the world's most vulnerable soils. They are also one of the world's main carbon reservoirs. By absorbing carbon dioxide from the air, storing the carbon and providing oxygen, tropical forests act as the world's thermostat, regulating temperatures and weather patterns. The loss of tropical forests contributes to between 12% and 15% of all greenhouse gas emissions each year – about the same percentage as all of the world's trains, aircraft and cars combined. In some tropical countries, such as Brazil and Indonesia, emissions from deforestation can be as high as 60%, which is higher than emissions from all other sources. Replanting and rehabilitating secondary forests around the world has tremendous potential for offsetting greenhouse gas emissions. Furthermore, rehabilitated forest lands can attract eco-tourists and offer a habitat for native forest wildlife.

Rapid Tree Growth in a Cleared Area of Rainforest

In order to gain maximum access to sunlight, nutrients and water, new trees grow quickly. This means that young plants take a much greater amount of carbon dioxide from the atmosphere for photosynthesis, the process by which energy from sunlight is used to produce the sugars that the plants need to grow. In the best conditions, new-growth vegetation could take up to 11 times more carbon dioxide as old-growth forests. However, the long established old-growth rainforests have locked away a vast quantity of carbon over the decades and centuries.

Endangered Species

Jaguars need large areas of habitat in order to survive. In Brazil, they may have home ranges as large as 142 square kilometres.

They are formidable hunters. In fact, the jaguar's common name means 'the beast that kills its prey with one bound' in Indian traditions. Prey, which is both large and small animals, including tapir, peccary, birds and fish, will be stalked or ambushed, then dragged to cover. They will even eat snakes, turtles, porcupines and caiman.

Females raise two cubs that stay with their mother for about two years. Although jaguars are primarily thought to be nocturnal, they are actually crepuscular (mostly active around dawn and dusk), but can also be active during the day.

The rainforests of the Amazon Basin have the largest numbers of jaguars. Jaguars like water and are very good swimmers and are often found near rivers. They are also good climbers. In captivity, jaguars can live over 30 years, but in the wild they are unlikely to even reach half this age.

It has been estimated that jaguars are now only found in 46% of their former range. In the 1960s and 70s, 18 000 jaguars were killed per year for their fur, which caused a devastating decline in their population. As farms and ranches encroach further and further into wild jaguar habitat, these big cats will sometimes prey on domestic farm animals so are often shot by farmers and ranchers.

Habitat destruction means that jaguar populations are becoming increasingly isolated and are, therefore, more vulnerable. There is no legal protection of jaguars in Ecuador and the hunting of 'problem animals' is still allowed in Brazil, Guatemala, Mexico and Costa Rica.

Malaysia and Indonesia (the island of Borneo) were covered in tropical rainforest with a wide biodiversity. Much of the lower lying areas have been cleared for palm oil plantations. Palm oil is used in cosmetics and many foodstuffs and is one of their major exports. The two countries produce about 85% of world supplies. The trade brings in a lot of foreign exchange and provides thousands of jobs. Plantations need the forest to be completely cleared. Trees of commercial worth are taken and the rest are burned. The survival of the orang-utan is seriously threatened as they are native to the lowland forest. Fragmentation of the forest means they do not have enough territory. Palm oil production is predicted to massively increase by 2020.

Conservation

There are many organisations which work to conserve the rainforests. They have developed a range of different conservation strategies to benefit the whole world, which also take into account the needs of local people and economic issues in the countries where the rainforests are found.

Organisation A

We are an NGO that works with local NGO conservation groups rather than governments. With donations from people around the world, we buy rainforest land to create reserves, which are passed on to the local groups.

We give scientific or technical advice and employ 'Rangers' who have a variety of roles: monitoring their area; keeping trails clear; maintaining fences; carrying out repair work after fires or storms; growing and planting trees; giving education talks; leading tourist walks; and assisting researchers. The presence of these Rangers deters poaching and other illegal activity and they earn a wage, which gives them independence and status within their community.

Our projects are small-scale but sustainable. We aim to protect the world's most biologically important and threatened habitats acre by acre.

Since being founded in 1989, we have funded partner organisations around the world to create reserves and give permanent protection to habitats and wildlife.

Organisation B

We are an NGO involved in conservation work. We work with governments, providing expertise and knowledge to help maintain protected areas that they have created. A typical scheme is working with individual native tribes in their own rainforest territory.

Donations from around the world provide boats, radios, fuel, border control training and aerial surveys. Surveillance has stopped illegal mining and mapped locations vulnerable to logging and fishing.

We are setting up non-timber, sustainable businesses with nuts, fruit, honey and essential oils. These provide jobs in harvesting, processing and transport.

We are dedicated to managing the things that we can control. We want societies to responsibly and sustainably care for nature and our global biodiversity, for the well-being of all humanity.

Answers

TOPIC-BASED QUESTIONS

Pages 148–153: The Challenge of Natural Hazards

1. Pacific Ring of Fire **[1]**
2. **Any suitable answer, e.g.** South American plate and African plate; Eurasian and North American plate **[2]**
3. **Any suitable answer, e.g.** Pacific plate and North American plate; Nazca plate and South American plate **[2]**
4. **Any suitable answer, e.g.** Tohoku, Japan (2011); Haiti (2010); New Zealand (2011); Sichuan, China (2008); Northridge, USA (1994); Nepal (2015) **[2]**
5. The focus is the point underground where the earthquake originates **[1]**, while the epicentre is the point on the surface directly above the focus. **[1]**
6. **Any suitable answer with reference to:** Plate boundaries; fault lines; measurement of fault movements; use of satellite imagery; frequency of previous earthquakes; tiltmeters; measuring gas release; unusual animal behaviour; difficulty of accurate prediction of timings. **[6]**
7. **Any suitable answer, e.g.** Water; tinned food; wind-up radio; wind-up torch; batteries; first-aid kit; survival blanket; clothing; solar light. **[Answer should explain how three items can be helpful. Up to 6 marks]**
8. **a)** False **[1]** **b)** True **[1]** **c)** True **[1]** **d)** False **[1]** **e)** False **[1]**
9. **Any suitable answer with a named example and impacts, e.g.** Tohoku, Japan (2011) or Asian Boxing Day tsunami (2004). **[1]** Impacts: large expanse of coast affected; worse closer to epicentre of earthquake; fatalities; loss of settlements and roads; extreme flooding; damage to ships; effect on nuclear reactor at Fukushima (2011); effect on water supplies. **[Up to 5 marks]**
10. **Any suitable answers, e.g.** Homelessness; buildings destroyed; unemployment; disruption to transport; shock and trauma; fatalities; pressure on medical care; migration; food; water; power supplies. **[Up to 6 marks]**
11. **Any suitable answer, e.g.** Merapi, Indonesia (2010); Eyjafjallajokull, Iceland (2010); Mt St Helens, USA (1980); Vesuvius, Italy (1944).
12. **a)** True **[1]** **b)** True **[1]** **c)** True **[1]** **d)** True **[1]** **e)** False **[1]**
13. D – Wind speed **[1]**
14. **a) Any suitable answer making the key point that the storms are not inland or along the Equator, but over warm ocean water between 5° and 20° north or south.** Pacific Ocean: east Pacific but mainly in the west Pacific near the Philippines and north-east Australia; north Atlantic; Indian Ocean (north and south of the Equator); rare in south Atlantic **[4]**
    b) Areas with higher sea surface temperatures tend to have the most tropical storms **[1]** but this is not the case along the Equator **[1]**. They do not form over land **[1]** and rapidly lose intensity over land because they lose their supply of warm, moist air. **[1]**
15. **Any suitable answer to show how the authorities adopted proactive strategies, e.g.** Inertia tackled; incentives successfully offered to encourage evacuation; as a result, when Typhoon Hagupit (also Category 5) hit the Philippines, only 18 people were killed. **[2]**
16. **Any suitable answer, e.g.** The prevailing winds are westerlies from the Atlantic Ocean, creating a maritime influence **[1]** bringing relatively warm, moist air from the North Atlantic Drift or the Gulf Stream. **[1]**
17. Relatively dry continental air from land masses to the south and east **[1]**. In winter, such air can be extremely cold, but in summer, very hot **[1]**.
18. It is in the middle of the Pacific Ocean, a very long way from large urban or industrial areas **[1]**. This means that there is no bias in the measurements. **[1]**
19. **Any suitable answers, e.g.** Pre-1850: largely natural causes (with examples and mentioning type of evidence). **[4]** Post-1850: a combination of natural and human causes, with human causes becoming increasingly significant (with examples, mentioning type of evidence and the hockey stick graph). **[4]**

Pages 154–155: The Living World

1. **Any suitable answer, e.g.** Deforestation in tropical rainforests **[1]**, as the removal of trees destroys the habitats of countless organisms **[1]**, thus leading to potential imbalance **[1]**.
2. **Any two from:** Northern Atlantic Ocean; northern Pacific Ocean; Australia; off the coast of southern South America; southern Africa; south-western Indian Ocean. **[2]**
3. The Mojave Desert / Sonoran Desert / Chihuahuan Desert / Nevada / Arizona / California / Western Desert **[1]**
4. Tourism that uses the beauty of the environment itself as the sole attraction **[1]** and helps to ensure the protection of the environment **[1]**.
5. **Any two from:** Water is needed for swimming pools, showers **[1]** and for the watering of golf courses **[1]** as well as for hotels, cooking and cleaning. **[1]**
6. Extreme tourism **[1]**
7. Absorbing nutrients **[1]** from the leaf litter **[1]**
8. Artificially watering farmland **[1]**
9. **Any suitable answer which gives at least one example and three well-developed statements, e.g.** Mining copper in the Western Desert of the USA creates many jobs and profits for companies **[2]**; using drip irrigation to grow crops such as fruit and nuts **[2]**; hydroelectric schemes generate energy, creating jobs and profits **[2]**; building settlements for retirement migrants brings economic development **[2]**; tourism in casinos and through the landscape creates jobs and profits **[2] [Up to 2 marks only if no example is given]**
10. The gradual process whereby a place becomes a desert **[1]**
11. **Any suitable answer with three well-developed statements, e.g.** Afforestation helps bind soil together **[1]**; terraced fields provide flat farmland **[1]**; bunds collect soil, preventing erosion **[1]**; gabions in rivers trap sediment, creating soil **[1]** **[The first statement gets 2 marks; each thereafter gets 1. Lists get automatic half marks.]**
12. A – The Sahel region of central Africa **[1]**
13. Methods that are cheap **[1]** and accessible for local people **[1]**.

Pages 156–161: Physical Landscapes in the UK

1. Angular or sharp-edged pieces of rock on the surface (scree) **[1]**; rock faces have jagged tops **[1]**.
2. **Any suitable answer, e.g.** Abrasion occurs when a glacier has material / debris / load at its base **[1]** that can erode rock as the glacier moves over it **[1]**; friction causes the ice to melt at the base of the glacier but the material / debris at the base is still able to abrade **[1]** before it is washed out **[1]** and replaced.
3. As it moves across the bedrock, meltwater from the edge of the glacier enters cracks and weaknesses **[1]**. It then re-freezes and expands **[1]**, putting pressure on the rock, which eventually breaks off **[1]** and can be carried away.
4. **Any suitable answer, e.g.** Harder rock is more resistant to erosion so will wear away less easily **[1]**. If the bedrock is less hard than the material at the base of a glacier, it will be eroded **[1]**. If it is harder than the particles at the base of the glacier, it will not be eroded **[1]** but will wear away the particles themselves **[1]**. **[Up to 3 marks]**
5. Crescent **[1]**
6. Arêtes result from the steepening of corrie back walls **[1]** to create a ridge between adjacent ones **[1]**.
7. **Any two from:** Animals can tolerate colder or wetter weather than crops; a less fertile soil is needed for grazing than crops; sheep can live on steeper slopes than will be able to grow crops. **[2]**
8. **Any suitable answer, e.g.** Deposits of minerals like iron, lead, gold were small **[1]** or not very high quality **[1]** or broken up **[1]** so difficult to mine **[1]**. Mines were small and quickly exploited **[1]** or deposits were exhausted **[1]**. Competition from other regions or countries **[1]**. **[Up to 3 marks]**
9. **Any suitable answer, e.g.** Diversification is different use of farmland and buildings **[1]**; former barns or cottages can be made into holiday cottages **[1]**; fields can be used as campsites **[1]**; the farmer may offer bed and breakfast accommodation **[1]**; farm tours or visits **[1]**, especially at lambing time **[1]** can take place. **[Up to 3 marks]**

10. **Any suitable answer with the main focus on erosion or disturbance, e.g.** Mountain biking, walking and skiing remove surface layers of soil / remove vegetation **[1]**, leading to soil erosion **[1]** or loss of vegetation species **[1]**. Rain can carry the loose material into watercourses **[1]**. Water sports can cause waves **[1]** that erode river or lake sides **[1]**. **[Up to 3 marks]**
11. **Any suitable answer with comparisons made, e.g.** On the valley floor soils will be thicker/deeper **[1]** because it is flat **[1]** compared to the sides, which can be very steep and so soil is not stable **[1]**. They will be more fertile **[1]** because streams wash sediments **[1]** from the valley sides. Valley floor soils will be wetter **[1]** because water cannot drain away easily **[1]**. **[Up to 4 marks]**
12. a) Hard **[1]**
    b) Rock armour / rip-rap **[1]**; sea wall **[1]**
    c) **Any suitable answer which explains how the structures work and what is being achieved, e.g.** Rip-rap is absorbing wave energy **[1]** so preventing erosion of the area behind the beach **[1]**. The sea wall is reflecting wave energy **[1]** so keeping the buildings and roadway safe **[1]**.
13. Slumping **[1]**
14. **Any two from:** Relatively low cost compared to other strategies; increases beach size for aesthetic; or recreational use. **[2]**
15. **Any suitable answer with two linked points, e.g.** It will protect **[1]** the cliff/cliff foot so prevent further erosion **[1]** / collapse of the cliff **[1]**. It will provide a base for more weathered material to build on **[1]** so the cliff face will experience less weathering over time **[1]**. **[Up to 2 marks]**
16. a) **Any suitable answer, e.g.** Plant roots weaken rocks as they grow in joints, bedding planes or cracks **[1]**. Decomposing plants release acid that could chemically weather rocks **[1]**. Surface vegetation can protect against damage/slow down attack by the weather **[1]**. **[Up to 2 marks]**
    b) **Any suitable answer, e.g.** Hydraulic action: waves trap air in cracks, joints and bedding planes **[1]**. This forces the air against the rocks at great pressure causing weakness/breaking **[1]**. Abrasion: debris carried in suspension or traction **[1]** scours / scrapes away the rocks **[1]**.
    Or Particles carried by waves are hurled against the rocks **[1]** and attack them, particularly at points of weakness **[1]**. **[Up to 2 marks]**
17. a) **Any suitable answer giving a named area and mentioning the physical reason for the protection and the human context, e.g.** Glacial till cliffs along the Yorkshire coast **[1]** are easily weathered and are eroding between 2 and 3 metres per year **[1]**. The B1242 road, connecting east coast settlements, was at risk **[1]**.
    b) **Any suitable answer for the same example, e.g.** Two rock groynes of Norwegian granite **[1]** were constructed at Mappleton to trap sand **[1]** and steepen the beach **[1]**. **[Up to 2 marks]**
18. More bedload means more friction **[1]** so the water needs more energy to move **[1]** and so will erode less.
    Or The bedload will protect the channel bed **[1]** so the water will not be in contact **[1]** and so will erode less.
19. **Any suitable answer about change to the channel form, not valley sides or rainstorms, e.g.** A tributary stream or river joining/at a confluence **[1]** means there is more water in the channel at that point **[1]**, which could cause more hydraulic action **[1]** or entrain more material **[1]**. A sudden steepening of the ground **[1]** means less energy is required for moving water **[1]** so more will be available for hydraulic erosion **[1]**. Faster water can entrain more **[1]**. **[Up to 4 marks]**
20. **Any suitable answer, e.g.** Floodplains are made from material deposited when a river floods / goes overbank **[1]**. This is left behind when water levels fall / the flood recedes **[1]**. There is also material from meander inner bends **[1]** that remains as meanders change/develop/move on the floodplain **[1]**.
21. Faults are breaks in rocks **[1]** that can expose different bands/resistance/hardness of rock **[1]**. Softer/less resistant rocks will be weathered and eroded **[1]** so, over time, rivers will meet a sudden change/step/vertical drop **[1]**.
22. **Any suitable answer that provides specific information to identify the landform and gives a sense of place, e.g.**
    Name of river: Hardraw Beck, tributary of river Ure.
    Name of feature: Hardraw Force.
    Location: Hardraw hamlet near Hawes (close to the foot of Buttertubs Pass) in the Yorkshire Dales.
    Key features: 30-metre drop, which is England's longest unbroken fall. The water falls over carboniferous limestone that overlies sandstone. The plunge pool is on shale. There is a rocky gorge in front of the waterfall called Hardraw Scar.
    [A maximum of 2 marks for location and 3 for features. Up to 4 marks in total]
23. **Any suitable answer with four points that cover the process (it is important to have the change to the channel occurring at flood/high discharge), e.g.** Water flows sinuously in a channel / moves from one side to the other **[1]**. Where it hits the bank, there will be greater erosion **[1]**. Eroded material is carried to the opposite bank so that bends are made **[1]**. The bends get bigger/more exaggerated **[1]** with reduced land between them. In times of flood/high discharge **[1]** water may break through, creating a new channel and leaving the meander cut off as an ox-bow lake. **[Up to 4 marks]**
24. **Positive:** Embankments can be made from material from dredged rivers **[1]** or excavations for industrial areas **[1]**, so their cost is low **[1]**. They will quickly have vegetation/plants growing **[1]** so will blend into the landscape **[1]**. As embankments are at the sides of a river, they do not affect the shape of the channel **[1]** so will not change the flow of the river / cause more problems **[1]**. **[Up to 2 marks]**
    Negative: Plant roots / burrowing animals can break them up **[1]** so making the embankment unstable/unreliable **[1]**. The material is permeable / allows water in / can get saturated **[1]** so can become unstable / let water through **[1]**. High floods can overtop the embankments **[1]**; flood water can get trapped on the side away from the river **[1]** so the embankment could slow down recovery from the flood **[1]**. **[Up to 2 marks]**

Pages 162–163: Urban Issues and Challenges

1. A city with a population in excess of 1 million people **[1]**
2. A city with a population in excess of 10 million people **[1]**
3. Urban sprawl **[1]**
4. a) Pull **[1]** b) Push **[1]** c) Push **[1]** d) Pull **[1]**
5. **Any suitable answers, e.g.** The creation of Queen Elizabeth Park; new housing; new sports facilities; the cleaning up of old industrial areas; improved transport systems **[Up to 4 marks]**
6. **Any suitable answers, e.g.** Sustainable housing; exploitation of renewable energy, such as the use of solar panels; recycling schemes; cycleways; good bus links; integrated transport system; available jobs **[Up to 6 marks]**
7. **Any suitable answers, e.g.** Public transport, like buses, trains and trams, means that people don't have to take journeys by car and this helps to reduce congestion on roads **[1]**. It also reduces levels of air pollution **[1]** and cuts carbon dioxide emissions (which contribute to the enhanced greenhouse effect and climate change **[1]**). However, public transport can be unreliable **[1]** and does not always go to places that people need to travel to **[1]**. In cities like London, schemes have been introduced to enable people to make short journeys by bike rather than by car **[1]**. Bikes create no pollution or greenhouse gas emissions **[1]**. However, some people are not confident enough to cycle **[1]**, and there are safety concerns about riding bikes along busy roads **[1]**. Some organisations encourage people to car share, meaning that fewer journeys are made **[1]**. This reduces levels of road congestion **[1]** and air pollution **[1]**. However, it is not always practical for people to car share as they may start or finish work at different times or have other commitments, such as picking up children from school **[1]**. **[Up to 5 marks if points are evaluated]**

Pages 164–165: The Changing Economic World

1. **Any three from:** Better healthcare; lower infant mortality; better diet; better water supplies **[3]**
2. **Any three from:** Contraception; financial reasons; fewer infant deaths; women pursuing careers **[3]**
3. **Any two from:** Tradition; high infant mortality; family security; economic reasons **[2]**
4. A **[1]**; C **[1]**
5. **Any suitable answers, e.g.** Teaching; doctors; insurance; banking; mechanics **[2]**
6. **Any suitable answers, e.g.** Any IT roles; research and development; software development **[2]**
7. **Any suitable answers, e.g.** Agricultural use of herbicides and pesticides; open cast mining; fracking **[2]**
8. **Any suitable answers, i.e.** Any forms of land, water or air pollution **[2]**
9. **Any suitable answers, e.g.** Excessive use of electricity; mining of rare earth metals for use in the manufacture of high-tech products like TV and computer screens **[2]**

10. a) **Any suitable answers, e.g.** Cheap labour; growing home market; fewer industrial laws and restrictions
    b) **Any suitable answers, e.g.** Development of new skills; positive contribution to GDP; extra tax revenue to spend on infrastructure; lead to attraction of other TNCs. **[Up to 5 marks]**
11. Intermediate technology is an appropriate form of technology for the context in which it is operating **[1]**. It is an improvement on existing simple technology **[1]** but does not necessarily make use of the most advanced technology available **[1]**. Credit can be given for naming specific examples you have studied **[1]**. **[Up to 2 marks for details of chosen example]**

Pages 166–169: The Challenge of Resource Management

1. A fuel such as coal, oil or gas, formed in the geological past from the remains of living organisms. **[1]**
2. **Any suitable answers, e.g.** The cost of building new power stations is likely to increase electricity costs to the consumer **[1]**. Other sources are cheaper **[1]**. Concern about the storage and disposal of waste **[1]**. Concerns about nuclear power stations being a target for terrorism **[1]**. **[Up to 3 marks]**
3. **Any suitable answers, e.g.** Farmers are stuck in the cycle of poverty **[1]**; lack of investment in agriculture due to lack of funds or government corruption **[1]**; climate change and drought **[1]**; civil war leads to displacement **[1]**; fluctuations in crop prices **[1]**; food wastage **[1]**; and poor storage and distribution networks **[1]**. **[Up to 4 marks]**
4. Food miles are the distance food is transported from the where it is produced **[1]** until it reaches the consumer **[1]**. A carbon footprint measures the total greenhouse gas emissions **[1]** caused directly and indirectly by an organisation / person / event / product and is measured in tonnes of carbon dioxide **[1]**.
5. **Any suitable answer, e.g.** Poor sanitation **[1]**, meaning people are exposed to cholera and typhoid and other waterborne diseases such as diarrhoea **[1]**. Many rivers, lakes and aquifers are drying up **[1]** or becoming too polluted to use **[1]**. **[Up to 3 marks]**
6. **Any suitable answers, e.g.** There are many different groups of countries: some, like those in the Middle East and North Africa are arid and have shortages of water **[1]**. Others like South Africa and India **[1]** have poor infrastructure and are not able to maintain supplies to their population **[1]**. Other countries like the UK and China **[1]** have large population densities **[1]** and water in some areas may be in short supply due to high demand **[1]**. **[Up to 4 marks]**
7. The range of different energy types used by a country. **[1]**
8. Place less reliance on imported fuels **[1]**, exploit their own fossil fuel reserves **[1]** and develop renewable sources of energy such as HEP and solar **[1]**.
9. **Any suitable answers, e.g.** Agriculture uses oil products to power farm machinery **[1]**, for the transport of goods and livestock **[1]** and in agricultural chemicals such as fertilisers and pesticides **[1]**. In recent years, agricultural goods like barley, maize and sugar cane have been used to make biofuels **[1]** that are used as a substitute for oil-based fuels **[1]**. Rising prices for oil means a higher price for biofuels and agricultural chemicals **[1]**, making food more expensive **[1]**. **[Up to 5 marks]**
10. **Any suitable answer, e.g.** Energy can be conserved through the use of certain types of construction materials, like mud and straw, that provide insulation **[1]**. The use of insulation in loft spaces and walls **[1]**, double-glazed windows **[1]** and larger windows in south-facing walls **[1]** are all methods by which energy can be conserved and energy costs reduced. **[Up to 2 marks]**
11. **Any suitable answer, e.g.** A 'sustainable transport system' is affordable for the population **[1]**, operates efficiently **[1]** and offers a choice of transport types (e.g. buses and trains **[1]**). The consumption of both non-renewable and renewable resources is limited **[1]**, and it reuses and recycles its components **[1]** to keep waste to a minimum **[1]**. Vehicle emissions are at low levels **[1]**, as is the level of noise **[1]**. **[Up to 3 marks]**
12. Biomass is the use of plants to produce energy **[1]**. Examples include the burning of wood **[1]** and the conversion of crops into biofuels such as ethanol **[1]**.
13. **Any suitable answer, e.g.** Technology has been introduced to increase efficiency in the use of fossil fuels. This includes power stations where clean coal technology has been introduced **[1]**. Technology has also been used to harness the power of renewable energy **[1]**. Cars have become more energy-efficient with the introduction of new engines that burn fuel more efficiently, reducing fuel usage **[1]**, and hybrid engines in cars like the Toyota Prius **[1]** and more aerodynamic designs **[1]**. Airliners have seen similar levels of efficiency introduced with new aircraft that use 25% less fuel than conventional airliners due to much more efficient engines **[1]** and the use of lighter materials in construction **[1]**. **[Up to 4 marks]**
14. Changing people's behaviour (in the use of energy) **[1]**
15. **Any suitable answer, e.g.** Greywater harvesting is a way of conserving water **[1]** and may involve recycling the water used in baths and showers **[1]** or recycling rainwater from roofs **[1]**, and using it for non-drinking purposes such flushing toilets **[1]**. **[Up to 3 marks]**
16. **Any suitable advantages, e.g.** The quality of groundwater is often very good since the soil and rocks through which the water flows help to filter out harmful components; groundwater responds slowly to changes in rainfall, so is often available during dry periods when surface resources have dried up; extracting groundwater can be quite inexpensive and does not require much technology (e.g. it can drawn from wells) so this makes it particularly important in LICs. **[Up to 2 marks]**
    Any suitable disadvantages, e.g. Using groundwater is not always sustainable and is vulnerable to over-exploitation; subsidence may occur if it is over-pumped; decline in the water table; groundwater resources take longer to replenish than surface resources. **[Up to 2 marks]**

Page 170: Geographical Applications

1. Answers will vary. **[Up to 9 marks + 3 for SPaG]**

PRACTICE EXAM PAPERS

Extended Response Questions

When your exam papers are marked, the total mark you are given for each extended response question (usually worth 4, 6 or 9 marks) will depend on the overall quality of your answer, including:

- the level of knowledge and understanding you demonstrate
- the level of accuracy, including correct use of specialist terms
- how well you use evidence/data/examples to support your ideas
- how well-developed your answer is:
  - o Is it a balanced discussion/does it consider a range of viewpoints?
  - o Is the meaning clear and easy to understand?
  - o Is it structured in a logical way that is easy to follow with a clear conclusion?

Don't be misled into thinking that spelling, grammar and punctuation is only important in the questions marked with **[+ 3 SPaG]**. Good grammar is essential in all written answers to ensure that your meaning is clear and your answer is not misunderstood. A good answer will always consist of well-developed sentences that use connectives.

For the purpose of this book, to help you mark your own answers, the key points that should be addressed in each extended response answer have been identified. However, it is important to remember that you will only be awarded full marks in the exam if these points are communicated in a clear, accurate and well-developed way (as outlined above). An example marks table that illustrates this has been included for Paper 1 Question 01.6. A similar scale will be applied to all 4, 6 and 9 mark questions in the exam.

Spelling, Punctuation and Grammar (SPaG)

The cover sheet of each exam paper identifies which questions will be used to specifically assess the accuracy of your spelling, punctuation and grammar. This is also shown by the marks in brackets alongside the relevant questions. For each of these questions, three marks are allocated for SPaG as follows:

- **High performance – 3 marks**
  - o spelling and punctuation is consistently accurate
  - o well-written with excellent use of grammar, so meaning is always clear
  - o a wide range of specialist terms are used appropriately
- **Intermediate performance – 2 marks**
  - o spelling and punctuation is considerably accurate
  - o good use of grammar, so meaning is generally clear
  - o a good range of specialist terms are used appropriately
- **Threshold performance – 1 mark**
  - o spelling and punctuation is reasonably accurate
  - o reasonable use of grammar (overall, any errors do not significantly hinder understanding of the answer)
  - o a limited range of specialist terms are used appropriately
- **No marks awarded – 0 marks**
  - o no answer has been given
  - o the answer does not relate to the question
  - o spelling, punctuation and grammar does not reach the threshold level (errors mean that the answer cannot be properly understood)

Paper 1: Pages 171–188

01.1 B **[1]**; C **[1]**

01.2 North American; Pacific **[1]**

01.3 **Any one from:** Magnitude refers to the strength/energy of an earthquake **[1]**; is measured on the Richter scale **[1]**; **Plus, any one from:** 7.8 is a strong/powerful earthquake **[1]**; has potential to cause a lot of damage **[1]**

01.4 Landslide/landfall/rock fall **[1]**

01.5 **Any two from:** Roads or other transport / communication links could be blocked; preventing evacuation/rescue/ arrival of emergency services **[2]**
Or Farmland could be covered/destroyed/altered; so there will be a shortage of food or a fall in farmers' earnings **[2]**
Or Houses could be covered / made structurally unsafe; leading to injuries and/or fatalities / creating homelessness **[2]** **[It is important to give the impact of the landslide and the effect on life for 2 marks in each case]**

01.6 **Your answer must:**
- refer to the stimulus material and case studies, which should be named and located
- describe and explain both problems and benefits
- include a balance of problems and benefits
- demonstrate good use of SPaG.

Benefits to include:
- good soils, e.g. Mt. Etna has orange groves and specific varieties of apricot and blood orange are only found there; Mt. Rainier is famous for its cherry and apple orchards
- tourism, e.g. Mt. St. Helens
- scientific study
- geothermal energy, e.g. Iceland.

Problems to include:
- respiratory disease/illness, e.g. Mt. Pinatubo
- lack of willingness to invest in area
- evacuation difficulties, e.g. Montserrat
- damage to farming – crops and livestock
- damage to transport infrastructure, e.g. roads, rail links, bridges
- pollution of water supplies
- damage to property.

Example marks table (at each level, marks move up with increased detail):

1–3 marks	• Description of problems or benefits only. • Reliant on stimulus materials.
4–6 marks	• Both problems and benefits are mentioned, although imbalanced. • Attempts to explain problems/ benefits as well as describe them. • Use of examples.
7–9 marks	• Both problems and benefits are described and clearly explained (some imbalance is allowable). • Relevant examples given. • Good use of examples.

01.7 **Example:** December 2015 was the wettest for 105 years (rather than 'since 1910') **[1]**; The bridge is an old one and its collapse was news, suggesting that it had not happened before / not happened for a very long time **[1]**

01.8 Eye **[1]**

01.9 **Any two from:** Pressure is rapidly falling at B, but stable low at A **[1]**; there is heavy rain at B, but no rain at A **[1]**; there is thick cumulus / cumulonimbus cloud at B, but none / clear at A **[1]**; there are very strong / hurricane force winds at B, but it is calm at A **[1]**; temperatures are relatively low (e.g. 24°C) at B, but much warmer at A (e.g. 32°C) **[1]** **[Each point must refer to both Area A and Area B for the mark]**

01.10 **Glaciers and ice sheets:** are all retreating **[1]**; with increasing speed in recent times / since 1970 **[1]** **Or** Decrease in sea ice **[1]**; could gain credit for linking the changes to global warming – causing ice sheets to melt, leading to thermal expansion in the oceans **[1]**.
Sea level: has been rising since 1900 **[1]**; 10 cm in British Isles / 19 cm globally **[1]**
Environment: Spring is starting earlier / autumn starting later **[1]**; causing other changes, e.g. altered migration patterns of birds / earlier appearance of butterflies / plants flowering earlier **[1]**

01.11 **Your answer:**
- must include three or four well-developed sentences
- must be restricted to economic activity and negative effects (references to animals, etc. will not earn any marks)
- must state two specific effects of climate change
- must give an economic impact for each effect stated
- might include named places.

Example points: Increased stormy weather **[1]** will increase coastal erosion and loss of farmland **[1]**; Higher temperatures result in lower snowfall **[1]** and loss of ski resorts / income from winter sports **[1]**; Increased frequency of hurricanes leading to effects on tourism **[1]**, farming **[1]** and fishing in coastal areas **[1]**.

02.1 entirely **[1]**; south **[1]**; some **[1]**

02.2 Hot desert **[1]**

02.3 Three arrows must be added, pointing left to right, from green plant to caterpillar, caterpillar to bird and bird to cat **[1]** **[All three arrows are needed for the mark]**

02.4 Green plant **[1]**

02.5 Emergent **[1]**

02.6 Lack of light hinders the growth of plants / only the few species adapted to low light can survive **[1]**; different tree types/species in tropical rainforests have a different growth pattern, so leaves grow and are shed at different times throughout the year. **[1]**

02.7 Thick and leathery: able to withstand heat and humidity without collapse / getting flaccid **[1]**
Or Drip tip: to shed water quickly/keep leaf dry to prevent growth of algae on the surface **[1]**

02.8 **Your answer must:**
- consist of three or four well-developed sentences
- make reference to weather, soil and rivers for 6 marks (but not necessarily in equal amounts).

Example: Reduced evaporation from the trees leads to less cloud formation and reduced rainfall **[1]**; The soil is less fertile as leaf litter, which provides humus/nutrient/ goodness, is reduced in amount **[1]**; Loss of tree roots, which stabilise soil, means it easily washes away **[1]**; Loss of tree cover/interception leads to increased leaching and soil erosion **[1]**; Increased sediment washed into rivers can cause blockages / lead to flooding **[1]**; River banks can become less stable and prone to collapse **[1]**; Less interception leads to more surface run-off, reduced lag times and increases in flooding. **[1]**

02.9 **Your answer:**
- must consist of three or four well-developed sentences
- must focus on human activity (points relating to natural phenomena will not earn marks)
- must make reference to case studies (a maximum of 3 marks can be awarded if no specific locations and groups of people are mentioned)
- might make reference to housing; transport; agriculture / food supplies; economic activity, such as mining; physical and mental health; education
- include ideas of actual challenges and how they have been overcome.

03.1 Attrition; Waves crash bits of rock together and as they are moved on a beach they become smoother and rounder **[2]** (The word 'attrition' does not need to be used, providing there is a full and accurate description)

03.2 Slumping **[1]**

03.3 Clay / Boulder clay **[1]**

03.4 Sea arch **[1]** (or 'arch')

03.5 Stack: A **[1]**; Sandy beach: D **[1]**

03.6 **Any two from:** Swash is stronger than backwash / backwash weaker than swash; more material is brought onto a beach than is removed; surging or spilling type; usually low height (<1 metre); long wave length, up to 100 m between crests; long wave period, 6–8 per minute **[2]**

03.7 Ro-Wen near Barmouth in Wales; Spurn Point at the end of the Humber / East Yorkshire; Orford Ness in Suffolk. Any example but must match the location **[1]**. Formation will depend on example used. There should be reference to the origin of material **[1]** and direction of movement along the coast **[1]**. Action of longshore drift should be clear **[3]** and influence of river channel and/or waves in shaping or stopping growth of the end **[1]**.

04.1 Some of the load is large in size, so there must have been much more water at some point to transport them downstream. [1]

04.2 There is moss/vegetation on some of the rocks in the channel, so they must have been exposed for a long time. [1]

04.3 A sudden change of gradient in the downstream course of a river. [1]

04.4 **Your answer must:**
- cover both elements for 6 marks, with a maximum of 3 marks if only one aspect is covered
- include detail / fuller explanations for more marks
- include examples of waterfalls or their particular situation (rock types, etc.)

Waterfall formation:
- faults / fault lines; sides of glaciated valleys; result of tectonic uplift; result of fall in base level; differential weathering.

Associated features:
- plunge pool – hydraulic action of the swirling water at base of drop deepens
- overhang (not all waterfalls) – mainly associated with differential erosion on harder/softer rocks; hydraulic action behind the fall
- gorge (not all waterfalls) – cutting back of the fall as rock overhang loses support
- bands of rock, particularly where harder rock sits above softer rock, e.g. High Force.

04.5 32 cumecs **[1] [The unit must be included to earn the mark]**

04.6 3.5 hours **[1] [Allow answers between 3.5 and 3.75 hours]**

04.7 **Any four from:** Increased discharge will be able to erode the bed and banks **[1]**; and so widen the channel / increase the size of meanders / or steepen the banks by undercutting **[1]**; leading to bank collapse **[1]**; Any potholes in the river bed could be deepened **[1]**; It would have carried a lot more sediment, which will be deposited downstream **[1]**; possibly on the floodplain, making it bigger, or within the channel, changing the channel shape **[1]**; e.g. on meander inner bends **[1]**; or in the middle causing braiding or as levees outside the channel **[1] [The landscape effect must be linked to the change in discharge to earn the marks]**

05.1 Pyramidal peak [1]

05.2 Lateral moraine [1]

05.3 Frost shattering / freeze-thaw action, in which water enters cracks in the rock and freezes and expands, putting pressure on the rock so it eventually breaks away **[2] [The term 'frost shattering' / 'freeze-thaw action' does not need to be used, provided there is a full and accurate description]**

05.4 Ribbon lake [1]

05.5 The lake has reduced in length / is shallower **[1]** due to infilling by material washed from valley sides **[1]**. **[Allow 'smaller' if the description is clear]**

05.6 **Any two from:** Slopes are too steep to form soil **[1]**; very little flat or gently-sloping land **[1]**; soils on flat land may get waterlogged **[1]**; accessibility issues **[1]**. **[Accept any other sensible answer; answers must refer directly to the landscape shown in Figure 19]**

05.7 **Your answer must:**
- consist of three or four well-developed sentences
- refer to both figures (a maximum of 3 marks can be awarded if only one figure is mentioned; some description of both will earn 4 marks; more detail is needed for 5–6 marks).

Key points to include:
- The footpath has been worn away / eroded by walkers.
- On the flatter ground of Figure 20a, it has become waterlogged and a boardwalk has been put at the side to prevent further erosion.
- On the inclined ground in Figure 20b, the footpath is much wider and lower; there is gullying where rainwater has washed surface material away; it has spread sideways so that the vegetation has been worn away; boards have been put across to make a staircase.
- Both figures indicate overuse / pressure on the landscape.

Paper 2: Pages 189–204

01.1 **Any two from:** Slow increase in urban population 1960 to 2010; slow decline in rural population 1960 to 2010; crossover in 2007, whereby there are more urban dwellers than rural **[2]**

01.2 **Any two from:** Rural to urban migration; growth of manufacturing industry; climate change; high birth rates **[2]**

01.3 Very large cities **[1]**; with over 10 million inhabitants **[1]**

01.4 **Your answer:**
- must include four or five well-developed sentences
- must explain how each issue creates challenges
- might reference infrastructure; transport; healthcare; education.

Example: Investment in infrastructure, such as sanitation, is patchy **[1]**. As a result, waterborne diseases spread easily **[1]**. Transport links are poor, so businesses struggle to expand **[1]**. Lack of investment in healthcare and education **[1]**. Therefore, death rates are higher **[1]** and local people are not equipped to build a better life for themselves **[1]**.

01.5 A [1]

01.6 **Your answer must:**
- name a regeneration project
- include at least four well-developed sentences
- mention both social and economic improvements.

Example: The Liverpool One retail development in Liverpool's central business district **[1]** created a huge number of jobs, mostly in the retail and food service industries **[1]**. The project also helped to alleviate poverty in the city **[1]** as many of the jobs created were semi- or unskilled and, therefore, open to all **[1]**. New transport links have helped businesses develop **[1]**. New hotels and visitor attractions have attracted tourists and therefore new wealth into the city **[1]**.

01.7 A **[1]**; B **[1]**

01.8 **Your answer must:**
- include four or five highly-developed sentences
- use geographical vocabulary accurately
- include examples (1 mark will be lost if no examples are given)
- demonstrate good use of SPaG.

Example: The BRT **[1]**; in Curitiba, Brazil **[1]**; subsidises bus transport within the city **[1]**, therefore dissuading lower income workers from using cars for transport **[1]**. Tubular bus stops, where passengers pay before they get on **[1]** mean the buses spend less time idling **[1]**. Bus routes are distributed across the city and cross at terminuses **[1]**, which means that the entire population is well served **[1]**. Poorer inhabitants of the city can exchange recyclable materials or excess farm produce for tickets **[1]**.

02.1 The number of births per thousand **[1]** per year **[1]**.

02.2 **Any three from:** Death rates; infant mortality rates; life expectancy; people per doctor; literacy rates; Human Development Index; access to clean water **[3]**

02.3 Gross domestic product (GDP) measures wealth but does not describe how equally it is distributed **[1]**; GDP does not take into account whether the money created lots of jobs **[1]**; GDP is recorded for the population as a whole and hides possible gender inequality **[1]**; credit can be given for an example, such as an oil-rich country **[1]**.

02.4 **Any two from:** Foreign investment; foreign government aid; foreign non-governmental organisation (NGO) aid; using intermediate technologies; fair trade; debt relief **[2]**

02.5 Free and subsidised contraception **[1]** allows low income families to control their birth rate **[1]**. Sex education makes both men and women aware of the mechanics of reproduction **[1]**. Family planning advice means people are more likely to weigh up the benefits and drawbacks of a larger family **[1]**. By having fewer children, families have more money, meaning they can have a better standard of living **[1]**. **[Up to 4 marks]**

02.6 400 m [1]

02.7 **Any three from:** Near to transport links; close to skilled workforce; cheaper land further away from CBD; brownfield land; government support / tax breaks; close to research intensive universities **[3]**

02.8 **Your answer must:**
- include four or five highly-developed sentences
- use geographical vocabulary accurately
- include examples.

Example: The growth of service industries **[1]**, such as telephone banking in India **[1]**, has meant higher paid jobs for the population, which has improved the standard of living **[1]**; More people working in higher paid jobs means that they can spend money elsewhere **[1]**, which is known as the multiplier effect **[1]**; Improvements in education **[1]**

mean that workers in India have more transferable skills **[1]**. Higher paid jobs mean more tax income for India as a whole **[1]**, which can be spent on infrastructure, such as roads and sanitation **[1]**.

02.9 **Any two from:** Recycling of heat from engines; use of wood from sustainable sources; building on brownfield land; paying living wage; generating own energy from renewables **[2]**

03.1 B **[1]**

03.2 **Any two from:** Lack of infrastructure; lack of investment; less taxes; weak government; corruption; population growth is too rapid for the water resources of the country **[2]**

03.3 **Your answer must:**
- include four or five well-developed sentences
- use relevant geographical vocabulary accurately
- give relevant examples.

Example: The areas of highest demand in the UK – London and the South-East – are the areas where there is least available water **[1]**. The areas with least demand for water – the north and the west of the UK – are also the wettest and have surplus water **[1]**.To maintain water supplies, water is transferred from wetter areas to those drier areas with greatest demand **[1]**. For example, water is sent from Kielder Water in Northumberland **[1]** to cities like Leeds **[1]** using a system of pipelines, aqueducts and rivers **[1]**.

03.4 Wind and solar energy **[1]**

03.5 **Your answer must:**
- include three well-developed sentences
- use detail from the graph.

Example: Consumption of oil stood at about 250 million tonnes in 1994 **[1]**. It then grew rapidly **[1]** to approximately 1000 million tonnes in 2008 **[1]**. After a small decline in 2008–2009, oil consumption rose again to approximately 1100 million tonnes in 2012 **[1]**.

04.1 A **[1]**

04.2 The UK has a much higher intake at 3413 calories **[1]**, whereas Somalia's is 1695 calories **[1]**.

04.3 **Any two from:** Famine; under nutrition; soil erosion; rising prices; social unrest **[2]**

04.4 **Your answer must:**
- include four or five developed sentences
- use an example of a more developed country
- give examples of sustainable food production.

Example: In the UK, organic farming means less use of harmful pesticides **[1]**. Urban farming initiatives, such as raising honey bees, can make use of brownfield sites **[1]**. Eating sustainably caught fish, such as pollock **[1]**, can lead to the stabilisation of depleted fish stocks, such as cod **[1]**. Eating more local, seasonal produce can lead to a reduction in air miles **[1]** and therefore carbon emissions **[1]**.

05.1 B **[1]**

05.2 In the USA, water consumption is roughly 575 litres per day **[1]**, whereas in China this figure is roughly 85 litres per day **[1]**; credit given for data manipulation.

05.3 **Any two from:** Waterborne disease and water pollution; drop in food production; drop in industrial output; potential for conflict where demand exceeds supply **[2]**

05.4 **Your answer must:**
- include four or five developed sentences
- give examples of methods of water conservation.

Example: Water conservation schemes can help reduce household usage **[1]**. One example of this is the distribution of water butts **[1]**. Managing groundwater supplies more efficiently **[1]** can help lessen depletion of stocks **[1]**. Recycling water can reduce both industrial and domestic usage **[1]**. Using greywater recycled from household use can cut domestic water usage and bills **[1]**.

06.1 E **[1]**

06.2 **Any two from:** Lack of infrastructure; African countries are poorer overall; political instability (war, etc.); issues with climate (deserts); lack of engineering and technical know-how; debt repayments to other countries means less funds available **[2]**

06.3 **Any two from:** Exploration of difficult and environmentally sensitive areas; economic and environmental costs; negative impact on food production; industrial output may suffer; potential for conflict where demand exceeds supply **[2]**

06.4 **Your answer must:**
- include a named example
- include four or five well-developed sentences
- mention both positives and negatives, skewed either way (a maximum of 3 marks if it only mentions one).

Example: Extracting tar-sand oil in Alberta, Canada **[1]**, brings many jobs into the area because of the huge need for labour **[1]**. Exporting the oil overseas brings money into Canada's economy **[1]**. Extracting tar-sand has incredibly large financial costs **[1]** because a huge amount of machinery and people are needed in the process **[1]**. Tar-sand extraction is highly damaging to the environment as local habitats and species can be completely wiped out **[1]**.

Paper 3: Pages 205–217

01.1 B **[1]**; E **[1]**

01.2 **Your answer must:**
- demonstrate some data linkage, rather than lifting information straight from the source.

Any two from: All three rainforest-containing continents feature in the top 10; countries where the rate of loss has reduced are in South America; the trend, up or down, is not related to amount of clearance. **[2]**

01.3 **Your answer:**
- must consist of four or five well-developed sentences
- must make good use of the resources
- must give more than three valid reasons with explanations
- must use well-developed sentences
- must use relevant geographical vocabulary accurately
- might mention plant and animal elements of diversity
- is likely to reference: low latitudes, size of area and features of the typical rainforest climate; relatively low population densities for a long period of time and the low level of technology associated with the indigenous populations having little impact.

02.1 Mean temperatures have risen over time (irregularly) **[1]**

02.2 Accept answers between 1°C and 1.05°C **[1]** **[The units must be given. Do not accept –0.42 to +0.62]**

02.3 1930–1940/1930s **[1]**

02.4 **Your answer must:**
- consist of four or five well-developed sentences
- explain the link between deforestation and climate change, e.g. forests absorb carbon and so keep Earth's temperatures regular / regulate weather patterns; deforestation releases carbon into the atmosphere, causing temperatures to rise
- describe some of the evidence of climate change (temperature change, storms, etc.)
- include impacts of climate changes on the economy of the world shown by the map
- include evidence of richer and poorer countries, taken from the chart
- demonstrate that you know the location of some of the countries named in the chart
- focus on the relationship between forest and climate (not impact on soils, etc.)
- reach a conclusion about the links between rainforest destruction and economic information, e.g. poorer places (using named places and figures taken from the map and chart), suffer most and richer ones, evidenced by the UK, suffer least.

03.1 **Your answer must:**
- consist of four or five well-developed sentences
- discuss a range of viewpoints, emphasising the conflicts/differences
- discuss evidence from the resource sheet
- consider the long term versus the short term.

Possible approaches include:
- the suggestion that richer countries were just as responsible in the past for forest destruction as locals appear to be at present (demand for forest products, etc.); also, emissions from long-industrialised countries have created problems in the rainforest and for the world, but have created high living standards – countries in Latin America, Asia and Africa might want a similar living standard / to catch up to some extent.
- the suggestion that richer countries have choices over employment / purchase of goods / way of life, which is not a luxury available to poorer people; the attraction

of ranching, plantations, mining, logging, etc. is that it offers immediate reward, even though the forest suffers; it is difficult to take a long-term view when living in poverty; certified goods might carry a higher cost but do not necessarily create as many jobs as quickly.

03.2 **Your answer:**

- must relate the strategies to the resource material provided
- must clearly explain how the strategies apply to a number of issues at a range of scales: local/regional, national and global
- might include forest protection, preservation and regeneration, biodiversity, local economy, weather and climate (both organisations can contribute to each)
- must consider the weaknesses of each set of strategies
- must reach a conclusion about which organisation's strategies would be most effective
- must demonstrate good use of SPaG.

Strengths: both sets of strategies will help with general rainforest issues, such as deforestation, impact on water supplies, weather and climate; they both work internationally so will raise awareness of the situations; their donations must largely come from richer countries so their advertising might cause people to change their ways of life.
Organisation A strengths: forest land is actually purchased so long-term control is possible; small parcels and new trees could fill in fragmented areas; focus on the forest gives prominence to biodiversity (reference to plants and animals); control is given to local groups who will understand the specific needs and way of life of native people; removed from government control so possibly less easily influenced; work with researchers to deepen knowledge and understanding; leading tourist walks to expand understanding to people outside the area.
Organisation B strengths: work with governments, so might be able to have greater power and influence and possibly avoid conflict; working with indigenous people maintains their identity; uses local knowledge (indigenous people); creates jobs and potential revenue/incomes for places with fragile economies; specific non-timber products reduce the threat to trees; marketing of products brings the rainforest issues to an international audience, which will increase knowledge and understanding beyond rainforest regions/countries; modern technology is used effectively for the good of the people and territory.
Weaknesses: in both cases, the suggested weakness is likely to be about scale – the strategies can only affect a very small area; however, a large-scale impact is possible.
Organisation A weakness: lack of economic development.
Organisation B weakness: possible government manipulation.

04.1 Important to get the idea of a small number or part of a whole and that it is representative. Measuring or counting a fraction / small part or group which represents the whole place/population. **[1]**

04.2 Taking information from along a line drawn on / from a map or along a road / other feature that is linear. **[1]**

04.3 A random sample, e.g. asking every sixth person who passes irrespective of gender or age, would be appropriate because it is easily managed **[1]**. It is reliable because there is no bias or attempt to get the same number of people in different groups **[1]** and so will represent the population in the town at the time **[1]**. **[Must cover both elements for 2 marks]**

04.4 25% **[1]**

04.5 Need not be coloured but two areas should be clear and the key completed. **[1 mark for each.]**

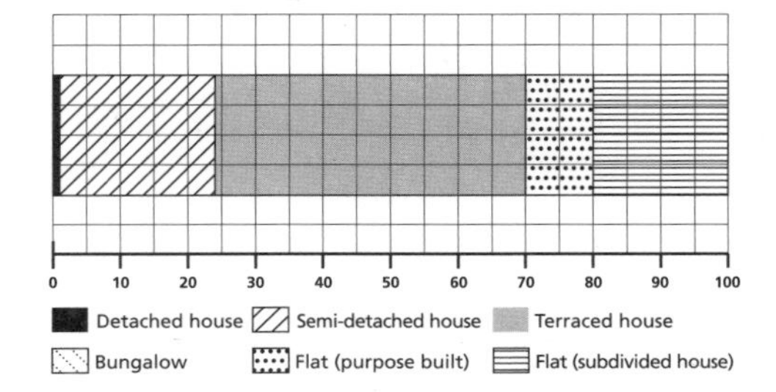

04.6 Terraced housing **[1]**

04.7 Trip line or desire line **[1]**

04.8 **T_1 is not the nearest town to village V so you are looking for a reason to explain this different behaviour to the norm / expected from the map. Any one from:** A larger town might have more choice **[1]**; there might be a faster road **[1]**; more frequent bus service **[1]**; faster public transport **[1]**

04.9 Vertical axis correctly drawn and labelled **[1]**; horizontal axis correctly drawn and labelled **[1]**; suggestion of plotted points **[1]**

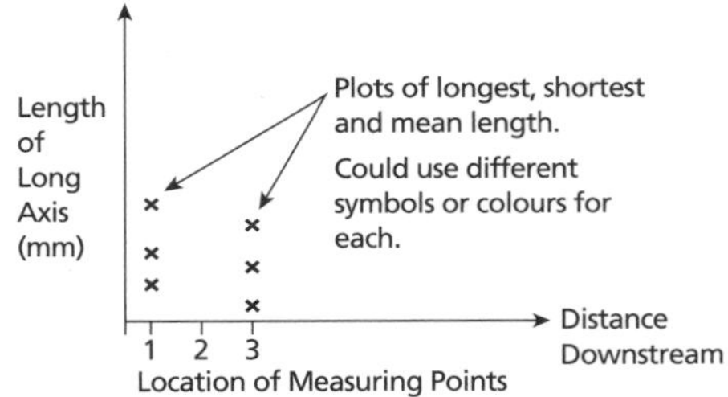

04.10 1 mark for each line **[2]**

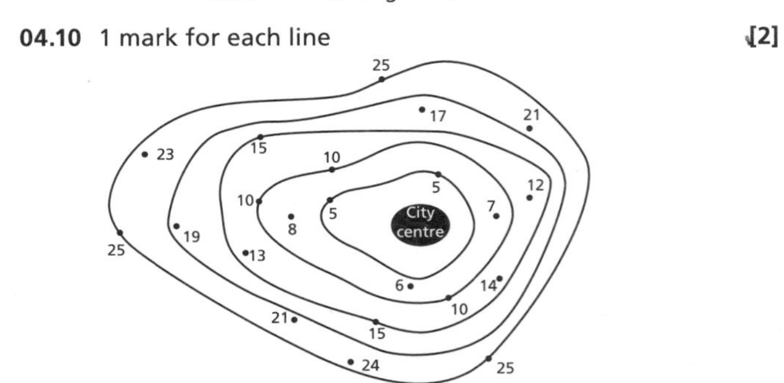

04.11 B **[1]**

05.1 **Equipment:** should be appropriate to the title of the enquiry. Description of use: for 2 marks, it should be possible to replicate the activity using the information. If lacking detail, 1 mark. **[2]**

05.2 **Your answer must:**

- be a presentation not a statistical method; name/ description of technique; there may be a sketch **[1]**; include an explanation of how data is made clear by using the method
- include reference to specific types of measurement needed, e.g. 'The width of the river in metres represented on a flow line which followed the course of the river. Scale used was ____ It was possible to instantly see how the river width varied along the course.' More detail accrues more marks. **[3]**

05.3 **Your answer must:**

- be analytical not presentation and must be relevant to the title; type of data used must be quoted, e.g. 'modal class of river bedload long axis' **[1]**
- include a statement of which data/figures were used **[1]**
- show understanding of the technique's use and applicability; no need for a result to be quoted but it might be included to reinforce the relevance
- include a comment about how the result/technique enabled clearer view of the situation **[1]**

05.4 Description only needed. Risk must relate to the title. Could be for any element of the work. e.g. traffic / road crossing **[1]**

05.5 Any risk relating to the answer given for 05.4 (could be for any element of the work) **[1]**. Should not repeat the description.

05.6 Conclusions must relate to any result quoted then, in turn, with the theories/concepts.

- Very general conclusions or application; only one element covered: **[1–3 marks]**
- Both elements with some understanding: **[4–6 marks]**
- Both elements and evidence of understanding of both the work done and how it fits with the underlying geography: **[7–9 marks]**